follow the water

FOLLOW

VARLEY LANG

THE WATER

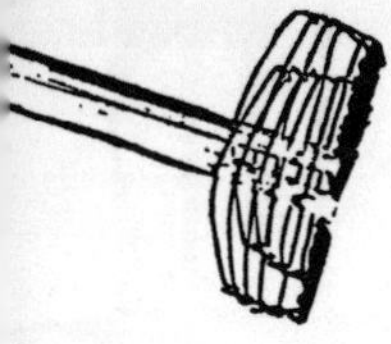

© JOHN F. BLAIR, PUBLISHER WINSTON-SALEM 1961

Library of Congress Catalog Card Number: 61-16637

Printed in the United States of America by Heritage Printers, Inc., Charlotte, N. C.

To
the watermen
of Maryland

Let the most absent-minded of men
be plunged in his deepest reveries—
stand that man on his legs, set his feet
going, and he will infallibly lead you
to water, if water there be in all
that region.

MOBY DICK

1 oystering 9

2 conservation 45

3 crabbing 69

4 clamming 95

5 fishing 109

6 hunting 133

7 boats 165

8 accidents 187

9 the character of watermen 201

Acknowledgments

The author wishes to express his appreciation to the many people who have helped him toward bringing his book to completion, and more specifically to the following: to his wife Helen, who sometimes culls for him and has had many opportunities of observing a tonger at work, a reproduction of whose carving of a tonger in Philippine mahogany appears on the title page; to Dr. Shields McIlwaine of the New York State College for Teachers, whose persistent prodding did much toward inducing him to get his book on paper; to H. Robins Hollyday, resourceful photographer of Easton, Maryland, for taking a number of pictures and giving permission to use others, reproductions of whose photographs appear on the jacket, the end-papers, and the title page and comprise Nos. 2, 5, 6, 7, 8, 13, 14, 15, 16, 17, 18, 19, and 20 in the body of the book; to John Moll of Oxford, Maryland, well-known Shore artist, whose map of the Eastern Shore appears on page 1; to David McKay Company, Inc., New York, for permission to quote the passage from Albert Klestadt, *The Sea Was Kind*, which appears on page 30; to The Mariners Museum, Newport News, Virginia, for permission to quote the passage from Marion V. Brewington, *Chesapeake Bay Log Canoes*, which appears on page 178; and finally, to Bill Mothershead, one of the few ship's riggers now active, photographs of whose models of sailing vessels comprise illustrations Nos. 21 and 22.

CHESTER R
QUEEN ANNE'S
KENT IS.
EASTERN BAY
WYE R.
TUNIS MILLS
TALBOT
ST. MICHAELS
EASTON
OXFORD
BAY
TILGHMAN IS.
CHOPTANK R.
DORCHESTER
MARY'S
WICOMICO
NANTICOKE R.
TANGIER SOUND
SOMERSET
Moll

In General

The Bay Counties of the Eastern Shore of Maryland are so laced with salt water that their coastlines resemble filigree, rather than solid littoral. Some areas, as in the southern half of Dorchester, are so inundated that it is difficult to distinguish land from water. Almost every large town on the Eastern Shore is located on navigable water, and the water is all salt. In many localities, there are fifty acres of water to every one of arable land. A list of islands within county waters and in the Bay adjacent would run into the dozens, and the word "neck" appears so often in county place names you sometimes wonder where the main can be, since a "neck" is a body of land surrounded on three sides by water.

In such a country, existence is likely to show amphibious habits. There is, in fact, constant ebb and flow from land to water and from water to land, either for pleasure or for work; or, more directly, for the purpose of getting food out of the water to bring back and eat.

Judging by the numbers of pleasure craft and party boats I see on the Bay and in county waters every summer, almost everyone on the Shore wets a line in the course of a season. On occasion, carpenters, farmers, boat-yard workers, store clerks, and ministers of the Gospel are also oystermen, crabbers, and commercial fishermen. Postal clerks and truck drivers will sometimes spend their vacations in a clam rig or with a drift-netter; and anyone in between jobs is likely to join a friend in the oyster fleet or lay a trot line for crabs, depending on the season. There are not many able-bodied men in the towns of St. Michaels or Oxford who have not handled oyster rakes, fished gill nets, or overhauled a trot line.

About ten years ago, during the oyster season, I went to a community in Talbot County called Whitman on the map, but known locally as Pot Pie. I was looking for a pair of nipper paws, but I found something of more interest.

Do you know what felts are? These are liners made of felt to be worn inside sea boots, and they are the same shape and size as the boot. They are warm and comfortable, so watermen tend to wear them when not working, as you would put on soft bedroom slippers.

It was a Sunday when I went to Pot Pie. It was not surprising that everyone I saw in the streets was wearing felts. Pot Pie is one of many watermen's villages. But the attendant at the gasoline station also wore felts. So did the man behind the counter in the grocery store, where watermen came of a Sunday for gossip and newspapers. When the church bell rang, the minister preceded his flock into the church. He was wearing felts, too, like all of his male parishioners. In a somewhat draughty building, he intended to be as comfortable as any. Like the gasoline attendant and the grocery man, he picked up the rakes whenever he got the chance. Was not Peter a fisherman, and a fisher of souls as well?

It is well known that ministers, in pursuit of their calling, frequently shift their place of residence. In 1955, the General Assembly of the State of Maryland passed a law which exempts any duly ordained minister of the Gospel from the residence requirements for a tonging license.

While ebb brings to the water many men, a flow often sets watermen ashore, stranded for the time being in the form of electrician's and plumber's helpers, carpenters and boat builders, house painters and sawmill hands. They are not completely at ease ashore, and their history shows that they will not remain there long. Once having wet the rakes or baited a trot line, men suffer from a virus which is often incurable. They are now subject to recurrent attacks whose symptoms vary. Some victims show a withdrawal pattern, a rather glassy look about the eyes, an absence of mind on the job. Others are aggressive and irritable, with a marked disinclination to take orders of any kind from anyone, punctuated, in more extreme cases, with outbursts of profanity and an occasional swing at the nose of anyone insisting upon cooperation. Both the withdrawal and aggression syndromes are sure indications that the victims of the virus are not long for this land. They will soon be back on the water.

Just as there are landsmen on shore who never set foot in a boat if they can help it, so there are watermen who never work ashore unless forced to do so. They spend nearly all of their adult years as professional watermen.

It is with these people and their jobs that the following pages largely deal. I intend to come ashore from time to time for various reasons, but I will not linger for long. I, too, have wetted a pair of rakes and baited a trot line.

oystering

1

September

Every year, toward the end of August, a feeling of change is in the air. In any community on the water you see men walking down those streets that lead to the harbor, and they will have on their shoulders two slender pieces of wood held together one-fourth of the way from the rake ends by a wooden pin. If you are a waterman, you will know that these twenty- to thirty-foot lengths of wood are tongs for taking oysters.

In the barber shop in St. Michaels, something new has been added. Alongside the wooden bailing scoops made by the barber in his spare time, now appear dogwood pins for holding rakes together. And the talk in the shop is about two hundred years old: it concerns oysters, oyster rakes, culling boards, culling hammers, the possible price per bushel of oysters, long-range weather forecasts, grapnels, length of anchor line, the set on Tide Mill, the condition of oysters.

In Dorchester and Somerset, smithies are working day and night, welding teeth on scrapes and rakes, forging iron rings for dredging gear. In ports all over the Eastern Shore, skippers check the gear of bugeyes and skipjacks, repaint scroll and figureheads with bright, new colors, test sheets and halyards, overhaul canvas and standing rigging. The dredge boats will not begin their work until November, two months later than the tongers, but their preparations take longer.

At dawn of the first day of September, 1957, 4,804 tongers left hundreds of ports from Somerset to Kent and steered their craft to the oyster bars: 609 from the ports of Talbot County, 925 from Dorchester, 633 from Queen Anne's, and 784 from St. Mary's. These men, along with tongers from other counties as well, were about to catch the major part of 2,190,074 bushels of oysters produced in Maryland in the 1957-58 season.

Of the 4,804 tongers, only 1,530 worked from the Western Counties of the Bay, so those fellows over there are not going to tell us anything about tonging.

Each year and every year, oysters produce more food and more money than any other resource of the Chesapeake Bay.

Time

Wherever man has lived near water, from the beginning of his generations, he has caught and eaten mollusks, among them, oysters. Examination of the kitchen middens of shore-dwelling Indians has revealed that ninety per cent of their diet consisted of shellfish.

One day some years ago, I was nippering oysters on a stone bed at Tide Mill in shoal water. Frequently young oysters attach themselves to bottom stones, so the culling board soon becomes loaded with a mixture of stones and oysters. In culling the load, the oysters are knocked off from the stones, and the stones themselves are thrown overboard. While culling a board in this way, I was about to toss over a stone from which I had loosened a couple of oysters. Some quality of form and smoothness made me hesitate. I looked at the stone more carefully and saw at once that it was an artifact, an Indian axehead, of the kind used by a squaw for splitting bones to extract the marrow, and, what is more interesting, for breaking the lips of oysters and other bivalves. The head was damaged in one place where it had broken along the natural cleavage line of the rock from which it had been fashioned and had, for that reason, been discarded. Some disgusted squaw had hurled the broken tool into the shallow water.

My patch of oysters had been worked a long time before.

The first colonists in Virginia saw Indians paddle their dugout canoes to shoal water and rake up oysters. They used a forked stick carrying long tines and a woven basket. The white men were impressed by the abundance of oysters and the comparative ease with which they were taken by the Indians. In 1623, one of the hungry colonists wrote a letter to the Virginia Company in England in which he says:

> The most evident hope from altogether starving is Oysters, and for the easier getting of them I have agreed for a Canaw which will cost me 6 pounds.

So evident was this hope and so easily realized that

the use of Indian dugout canoes spread rapidly throughout tidewater Virginia. Before very long, oysters became a regular part of the diet of people living on or near the Chesapeake Bay.

By the middle of the eighteenth century, increase in population and the growing economic development of the Eastern Seaboard made professional fisheries in the Bay area a natural outcome. Tonging oysters for market rather than for personal consumption became an increasingly important industry after 1760. Within twenty-five years of that date, oysters were a well-known food throughout the tidewater. Taverners and innkeepers in particular found them a favorite dish with travelers. Oysters remained cheap and plentiful throughout the eighteenth and most of the nineteenth century, not only in America but in most of Europe as well. They were truly the poor man's food everywhere.

Charles Dickens, when he wanted in any of his novels to set off the character of a neighborhood in London at one stroke, simply stated that great heaps of oyster shells were to be seen near the houses. His readers were at once informed that the neighborhood described was a poverty-stricken one.

Ten years ago, I took a small jag of oysters to a buy boat lying off St. Michaels harbor. After I had sold them, I had a chat with the captain. Among other things, he talked about oysters. He insisted that they are poor man's food; that, if laborers have no money to spend, no oysters will be sold.

Even that long ago, the captain was thinking in terms of the past, as a glance at present store prices for oysters will confirm. Oysters at $1.39 a pint are hardly poor man's food. Unshucked oysters are even higher. It seems unlikely that the poor man is going to have much of a shell heap near his house.

The market is no longer a cheap bulk one; it is very nearly a luxury market. There are good reasons for this development, and I want to say more about that later. The point to make now is that there is a long tradition of professional oystering in the Bay, and that, with the exception of the use of gasoline engines instead of sail, the ways and means of catching oysters have not changed for two hundred years.

Tongers

The gear needed by a tonger in order to work efficiently is very simple; no complicated mechanisms are required.

He needs a seaworthy, open boat with a small cabin forward in which to keep a stove, his dinner box, and a couple of life preservers. He needs about three pairs of rakes, from sixteen to twenty-four feet in length. If he favors deep water, he may carry a pair of twenty-eights or thirties. Rakes are essentially the same tool as that employed by tongers two hundred years ago. They consist of two shafts or handles, rounded on the outsides to fit the palm of the hand. These shafts are made either of edge-grain fir or of Georgia pine. At exactly one-fourth of the distance from rake heads to end of shafts, a pin of dogwood is inserted to hold the shafts together like a pair of scissors. The portion of the shaft from the pin to the steel rake heads is called the "stride." It is so named because the length from pin to rake head determines how far apart the heads can be held on the bottom.

When the tonger rests the rake heads on the bottom,

he feels about until he strikes oysters with the heads. He then opens the shafts with his hands, one on each shaft; and, by so doing, he opens the rake heads on the bottom, exactly as you would open a pair of scissors by pushing the handles apart. He then rests the teeth of the parted heads firmly on the bottom and rakes together a pile of oysters. When the rakes are brought together, the raked oysters are now held in a basket, the steel teeth forming the bottom, and iron rods placed above the teeth on both shafts, the body. The two shafts are now together because the heads have been closed. The tonger lifts the closed shafts hand over hand until the pin appears above the water. He slides one hand toward the pin and lifts the heads from the water and over a culling board. As the heads cross the board, he opens the shafts slightly, thus parting the heads. What the heads hold falls onto the culling board.

The culling board he makes himself from planks, to which he adds a collar to keep the oysters on the board. It is placed thwartship and open at both ends to allow the culler to brush shell and small oysters overboard.

When the tonger culls, he uses a highly specialized tool, the culling hammer. It weighs one-quarter of a pound and is absolutely worthless for any other purpose. With this hammer he cleans his market-size oysters. He knocks off excess shell with the head, and small oysters with the blade.

Boat, rakes, culling board, and hammer: that is all the gear needed.

The tonger leaves his port in the early winter morning, ice and weather permitting, and chooses a place in the river or Bay where he has learned from long experience that oysters are likely to be found. His perennial hope is to find a "patch," that is, a place where the oysters have been worked over little or not at all for some time. Having located his ground by a sample rakeful or two, he anchors his boat astern, if he is a Talbot County man (Mud Turtle). If he is a Queen Anne's man (Yellow Belly), he will anchor from the samson post on the bow. The Mud Turtles have always anchored astern. They catch more oysters that way. The Yellow Bellies have made their grapnels fast at the bow time out of memory. They catch more oysters that way. This ancient distinction still provokes heated arguments which have not been resolved in generations.

I am a Mud Turtle myself, so let us say the tonger has anchored astern. He then stands up on the washboards of his boat and begins to tong either at the bow or at the stern, emptying his catch on the culling board, which, when full, he "culls off," he or his partner, if he has one. He rakes the bottom on both sides of his boat, a complete circuit, unless he has a mate, in which case each tongs a side.

Having made his circuit, the tonger then pays out line the length of his boat so as to get on new ground, makes fast, and goes to work again in the same manner. The wind, of course, moves his boat until the line tautens. If there is no wind, the tide will do the trick. If there is neither, the tonger takes up his grapnel and throws over a killick on a short line. A killick is a smooth piece of metal with no flukes. In other words, it will not dig into the bottom. The tonger then pushes his boat by poling with his tongs.

Assuming there is a breeze, however, let us suppose that the tonger has found oysters on the port side of his boat, and few, if any, to starboard. In that case, he shifts his anchor line from the center of the stern to the starboard stern post, thus sheering his boat to port and over the better ground.

If he works all day steadily and with intelligence on good ground, he will have "a pretty good little jag," say from six to ten bushels of oysters. If he has loafed and used poor judgment, he will have "a capful," and he had better think seriously about getting a job ashore that he can handle.

Having gotten a pretty good little jag, the tonger pulls in his grapnel and goes either to a buy boat or to an oyster house to sell his catch.

The Market

One of the principal markets for the tonger in the Chesapeake Bay is the buy boat. It is a power craft which hauls other cargo in summer, such as wheat or slag, and oysters during the season. In the days of sail tonging, the power buy boat performed a very important service. At the end of the working day, these vessels came to every individual log canoe over a wide area and collected the catch, thus saving the tongers a long sail in heavily laden boats.

Even today, however, the buy boats are useful. They pay the market price, and their loads gathered from workboats are then carried either to packers or to a wholesale fish market. Some boats run to Baltimore for the latter outlet.

The captain of a buy boat is competing for oysters

with the many packing houses located all over the Bay, and, in some cases, with trucks from Delaware and Pennsylvania which drive down to local wharves and buy oysters from the workboats. The trucks then transport immediately to Wilmington or Philadelphia, thus assuring raw bars and sea food markets a fresh daily supply.

The multiplicity of buyers means a fair price to the tonger, since each buyer is competing with the others. For this reason, I believe that buy boats should be encouraged, and that it is a good idea for oystermen to sell to the boats as well as to the packers. In fact, a trustworthy captain of a buy boat generally builds up a clientele, a string of twenty or thirty boats which can be depended upon to sell their catch to him. A single packing house, for example, dominating a large area of water and the sole outlet for a fleet of workboats, is subject to a nearly irrestible temptation—price cutting.

Let us say, then, that you have brought in your catch from Herring Island or Hambleton Hills. Just outside St. Michaels harbor lies a buy boat anchored on the lee shore. You run alongside and tie up. The captain passes the time of day, tells you the market price per bushel of oysters, and lowers from a boom an iron measure, exactly and precisely one and three-tenths of a United States bushel. This measure has holes in its bottom to drain water from the wet shells of oysters as it is filled. There is another bucket on the other side of the buy boat, so that two boats can unload or "put out" at once. You shovel your oysters into the bucket until it is filled. Whereupon the captain or one of his crew hauls it to the deck of the buy boat, cants the contents into the hatch, and sends back the bucket.

The moment that iron measure rests on the floor boards

or ceiling of a workboat, a conflict of interest begins. The buyer wants as many oysters in the bucket as possible; after all, he is paying for them. The oysterman has every intention of shoveling in as few oysters as he can; he has worked all day for those oysters. The law says the measure must be filled level to the brim.

The result of this tug and pull is that some tongers have acquired a high and polished state of proficiency in loading a tub. It is best to put the oysters in lightly, leaving plenty of air space between the shells. It is said that some Queen Anne's men could shovel an iron tub full of eggs without cracking one of them. I have seen such an expertly loaded tub strike accidentally against the side of the ship as it was being hauled aboard. The jar shakes the oysters down and leaves the tub about four-fifths full.

On one occasion, Captain Peter Voight was being admonished by the skipper of a buy boat.

"Fill 'em up, Mr. Pete, fill 'em up! Now you know that ain't a full measure, fill 'em up!"

Captain Pete said, "You want me to put my whole load in one bucket? Haul away."

When the tub had been hauled halfway up, Captain Pete held up his hand. "Hold her! Now," he said, "is she full or ain't she?"

"She's *level,* Mr. Pete, but you and I know she ain't *full.* I seen you shovel. Why, them oysters is hung so bad, ain't a thing but air in that tub!"

"Send her down!"

Captain Pete guided the bucket to the culling board and tipped the contents out. He then stood on top of the board with the shovel in his hands. He spat first into one palm, then into the other.

"Shovel," he said, "I'll show you how to shovel!"

From the culling board, he shoveled back every oyster he had just emptied from the tub. When he was through, that measure was so heaped up, oysters were falling over the top. The buy boat skipper never bothered him again.

Expert shoveling is not thought of by watermen as being dishonest; it is simply proficient, one of the many skills needed by a successful tonger.

Let us suppose that two tongers, working together in the same boat, a common practice, have had a good day. They have come in to the buy boat with nineteen bushels. They put out and, as they do, either the captain of the buy boat or a helper keeps tally or count on the number of bushels loaded into the measuring tub by the watermen. The system of tally keeping is remarkable for its simplicity. It is said that very primitive peoples have no conception of numbers over five. We seem to be in the same class.

When one bushel is emptied, the tally keeper makes a mark in his book, thus: /. When five have accumulated, his tally is 𝍸 . He then calls in a loud clear voice, "Tally!" The waterman shoveling away in the bottom of his boat answers antiphonally, "Tally!" The count continues as before until ten is reached (𝍸 𝍸), whereupon the tally keeper calls, "Tally Two!" and the oysterman replies in kind. When the last bushel is emptied, the tally is called, "Four and all!" That is, four bushels out of a tally, and the count ended, "and all."

Should the tonger want to know at any particular time during the shoveling just what his count is, he is told at once exactly where he stands. If his tally at that point happens to stand at six bushels, the reply to his question is "One to One!" That is, one bushel plus one tally. If his count happens to be twelve bushels, the reply is, "Two to Two!" two bushels plus two tallies. The final count of nineteen is, of course, four to three, four bushels plus three tallies.

In this way, no number larger than four is ever spoken. We do not have to puzzle our heads with such complex figures as twelve or sixteen. A Tasmanian or a Malay pigmy could learn this system and use it without too much trouble. We are a plain and simple folk. Just the same, when I first started to oyster, it took me a long time to find out just what the devil was going on with all these "Three to Two's" and "Two to One's" floating around like odds at a race track.

Having interpreted these cabalistic calls with ease and put out all his oysters, the tonger casts off bow and stern while the captain counts out his pay in the wheel house where he keeps the cash. No checks or anything newfangled like that, but real, green cash. Now both are happy: the oysterman is paid for his labor, and the captain anticipates fifty cents on the bushel as pay for transportation at delivery.

Why Not More?

I have often taken friends out with me in the fall of the year. They look at the oyster fleet happily raking away, the men calling to one another, joking and laughing. From two-thirty on of a fine afternoon, workboats start homeward, the tongers at ease in the sterns or standing cockily with one arm akimbo, a nice little jag on the floorboards and green cash at the end of the run. They whistle and sing, happy as larks.

At this point my friends always ask, "Why aren't there more of them?"

Well, it does look good in September and October. The days pass like a dream. It seems you have no sooner begun to tong than the time comes to quit. You get a decent return for your work. You come home, have a drink or two, and tuck into a big meal, and you've never tasted anything better in your life. Just before you drop off to sleep at night, your mind recalls images of quiet, wide waters, the wheeling of gulls and sparkle of sun. You sleep like a child, and you waken eager for the new day.

Nevertheless, my friends have asked the wrong question. It should read, "Why are there so many of them?"

When you consider what a professional waterman has to contend with, it is a great wonder there are any at all in these easy days of desk jobs and secure employment. Watermen work hard for their living in all kinds of weather and water, much of it unendurable by modern standards, some of it distinctly dangerous.

Perhaps I can illustrate. Two winters ago, I looked out of the kitchen window every morning for a week or more to see the same old ice, bank to bank. My workboat, which I call *Sammy* after a novel entitled *What Makes Sammy Run?* was locked in. The weather warmed up some after a while, and we got a heavy snowfall. I walked out on the ice and snow to the stake where I keep my boat moored, and I shoveled the snow out of the boat. Ordinarily, a man pumps his boat. I shoveled mine, and every tonger on the Shore was doing the same.

A week later rain and wind melted the ice, and I was ready to go, eager to go. I was getting hungry. Unfortunately, a calm night made ice again. The next night saw a brisk, cold norther and a million million stars. The wind opened the water once again, as I saw when I looked out of the kitchen window in the morning. The temperature stood at fourteen above zero by the thermometer on the front porch, but the water was open due to the persistent wind blowing right down Leeds Creek. After I had paddled the skiff out to the boat, the first thing I did was to check the thermometer fastened to one of the knees. It also stood at fourteen above zero. I said, "Jesus," and started the motor.

The wind was twenty miles an hour from the north. It is a known fact that at thirty-two degrees Fahrenheit, every mile an hour of wind has the effect of lowering the temperature of the air by one degree, as far as the effect upon the human body is concerned. You work out the arithmetic. All I knew was that it seemed uncommonly cold; so cold that by the time I reached the mouth of the creek I knew I would have to start tonging soon or freeze. I turned down river to the nearest bar on Second Point. As I turned, *Sammy* took the sea on her bow and threw spray aft over the cabin and onto the washboards. Where the spray hit, it froze at once. By the time I anchored, the entire boat was covered with slick glaze.

I put the twenties overboard and went to work. I had to be very careful with my footing because the washboards were by now well covered with ice. One bad skid, and I might be overboard, and all that river wanted was a man for breakfast. You could see that by the way he showed his white teeth.

When I made my first lick with the rakes and lifted the load, the shafts came up covered with ice. After a while, the rake shafts got so slippery I was unable to manage them. When I lifted them hand over hand with my rubber gloves, they slipped, and I wasn't about to take my gloves off. I needed those hands.

Finally, I came down from the boards and searched about in the cabin until I found a couple of pieces of old toweling and some fishing twine. I wrapped the rake shafts with the toweling and lashed that in place with the twine. Of course, the cloth froze at once, but it was rough, it afforded more traction than the hard and slippery wood. That did the trick.

I have never learned the waterman's art of nose blowing. It can be defined as the "free" or "airborne" blow. I have to use a handkerchief, and in cold weather I have to use it often. To save the trouble of constantly digging under my frozen skins and into the hip pocket, I put the handkerchief on a little shelf under the washboard, where it immediately froze. Every time I wanted to use it, I had to beat it out on the engine house to remove the ice crystals. There is no satisfaction at all in blowing your nose on a hunk of ice.

After what seemed a very long time, I got a board and I began to cull it. First, I had to take an old carpenter's hammer that I keep in the boat and break up the frozen mass of shell and oysters. As I was doing that, a workboat came roaring up at full speed and then slowed down alongside, and Captain Howard lifted a hand. I waved in return and tried to grin, without much success. Captain Howard drawled, "You been here all night?"

I had to laugh then. Here I thought I'd done a pretty hard-nose piece of tonging, and that was all I got by way of comment.

And that is all you ever will get. In the first place, it is hard to impress watermen under any circumstances. In the second, a rough bit of tonging is all in the day's work. You tong or else.

Many watermen work in worse weather than I do. When it gets really bad out there, when "it ain't fit," I tuck my tail between my legs and run.

The tonger always feels the pressure of time on his shoulders. If he loses a month or two of working days during the season, he can not expect to have an extension of time to make them up. When April the fifteenth rolls around, he hangs up his rakes: time lost is gone forever. And so much time is lost.

A northwester will send him ashore three days running. On the next day, a beautiful blue sky and a glorious sun will send him singing to work. The only trouble is that the wind is blowing up the river and a spring tide is running down it like a race horse. When the tonger anchors, the resultant of two forces keeps the boat neatly and constantly in the trough of the seas. He stands on the washboards and tries to balance and tong at the same time, up and down, up and down, until the damn boat nearly rolls the teeth out of his head. He quits. There is nothing else to do. And he loses time.

Snow, sleet, storm, and ice: and the worst of these is ice.

In the winter of 1917-18, both the county waters and those of the Bay proper froze so hard trucks could be driven over the surface. Tilghman Island tongers cut out spaces in the surface and tonged from the edges of the ice. Not many years later, a buy boat in St. Michaels harbor filled her hold with eight hundred bushels of oysters. She was ready the next morning in early December to carry her cargo to Oxford. She was unable to leave until the following March, due to ice.

Of course, under such severe circumstances, tongers do not work. Their boats are not ordinarily equipped with runners. But even overnight freezing can cause havoc. What you get then is window-pane ice about one-quarter of an inch thick that cuts like a razor. A workboat can easily break such ice, and the temptation is great to run through it to open water. Within two hundred yards of running, window-pane ice will pull the cotton out of the seams and score the planks at the waterline from cutwater to transom. A run of a mile will sink your boat. A Queen Anne's waterman lost his workboat in just that way three or four winters ago. His planks were cut through clean to the chine. Fortunately, she went down in shoal water.

The patrol boats of the Tidewater Fisheries Enforcement Officers are generally sheathed in copper. In pursuance of their inspection duties, they must patrol. When they leave harbor in St. Michaels, Tilghman's Island, and dozens of other ports on the Eastern Shore, watermen follow their wakes through the ice. This is a very important service, one of the many performed by the Tidewater Police.

The following letter, signed by thirty-one Neavitt watermen and printed in the *Easton Star-Democrat* for 6 January, 1961, illustrates both the help ordinarily given by the patrol and the endless trouble watermen are caused by ice:

> A few weeks ago, the Tidewater Fisheries Commission brought in [to Neavitt] one of several new boats purchased from the Chris Craft Corporation and removed the boat that had for years been used very satisfactorily in this area.... At the present time, this new boat is frozen in the ice and absolutely useless in the enforcement of the oyster law....

> The boat replaced in this transaction [*Vera Bee*] was sheathed in copper metal for ice and has for years broken through ice in the course of law enforcement. Due to this fact, the officers aboard have been able to do a very commendable job during freezing weather....
>
> To make the present situation more ridiculous, the officers stationed here in Neavitt were forced to take a 15 foot rowboat that we know to be over 25 years old and break through the ice in order to carry out their duties. The use of this old rowboat for this purpose undoubtedly endangered...the lives of the officers involved, and we feel that commendations should be extended to Milton Higgins and Weldon Bridges for their efforts....

It is evident that the thirty indignant signers of this letter have the conviction that a new low has been reached when the State Oyster Patrol has to ship in a rotten rowboat. What is more important, a brand new craft has been issued without any metal on her for breaking ice.

Why don't watermen sheathe their own boats in copper? The answer is that the job is very expensive. Furthermore, metal should not be applied to old wood, for the copper tends to rot the covered planks. In addition, the sheathing is heavy, thus slowing the boat and hampering her performance. Copper should be applied to new wood, preferably before the boat is launched after building.

In winter, I use oak planks at the waterline. But it is difficult to get green heart oak of the proper length and width.

Finally, the sun shines, the wind blows, and the ice breaks up. But even then your troubles are not over. There is a lot of loose ice about, and it can drift in packs until it closes in on you. When that happens, you have to run through ice to get back to port. You have no choice.

Our friends from Neavitt have had a hard winter, no doubt of that. Two of them were caught in pack ice just a few weeks ago. The two tongers were afraid to risk their boat and possibly their own lives by crushing through the loose ice, but they had to get through or spend a freezing night on the water. They rummaged around in the cabin of the boat and gathered every old jacket, coat, and sweater they could find. These they hung in such a way as to cover the waterline and so give the planks at least partial protection from the ice. After running a while, the two tongers saw that more protection would be needed, and so added their own clothing. It must have been a right chilly run. Still, they made it. Of course, all of the clothing was torn to shreds.

The winter of 1960-61 is making a vigorous effort to become one of the worst ice years on record. Immediately following a snowstorm on 10 December, the county waters froze and then gradually encroached upon the Bay proper.

When between eight and ten inches of ice had formed on the Miles River, the port of St. Michaels requested help from the coast guard. In the past, watermen had ice-skated down the river to meet the old ice-breaker, *The Latrobe*, for the men were glad of her help. On this occasion, their joy at seeing the ice-breaker was somewhat darkened by the lateness of the effort and the damage done to tonging boats in the process. The harbor ice was so hard that tremendous lateral pressure caused by the coast guard vessel in parting the ice severely strained most of the boats and broke the planks of two of them.

The ice-breaker has since been referred to as the "boat-breaker."

In early January, a week's thaw cleared the waters of ice for three or four days, after which they froze again. This time the entire Bay, from the Capes to the Delaware Canal, was covered with ice. It is now 8 February and the ice is still heavy over most of the Bay. Oystermen from Bosman walked out on the ice of Broad Creek and there cut holes in the ice in an attempt to tong oysters. But the ice was from eighteen inches to thirty inches thick, and the labor necessary to open a space sufficiently large for tonging was so arduous and time-wasting as to be scarcely worth while.

The Attractions

I hope you begin to see why there are comparatively few watermen. In fact, why does any man choose to burn in summer and freeze in winter?

Well, one thing is obvious at once. The waterman is his own boss. Not everyone is happy in that condition, but there are a few men everywhere who are never happy in any other; and among them are watermen. It is primarily for this reason that watermen are not long satisfied ashore. Very definitely, those who like the water are fond of its freedom. Of course, freedom is a relative matter. We are all bound by economic laws. Also, we carry within us our own limitations, some of them serious. But I know of few trades any freer than that of the crabber, the fisherman, or the tonger. He owns his own boat and gear. There is no one on God's earth to say he must go out at a certain time of day or wait for the whistle before quitting. If he doesn't like the weather, he can come in or stay out, depending on his own wishes and his own judgment.

There is still another attraction in this trade, one more subtle and seldom expressed, but not the less cogent for that. In our time, clear-cut issues scarcely exist, except in the platform speeches of politicians. Whom do we blame? Who is to be praised? Blame is shifted from the shoulders of responsible individuals to irresponsible generalities, such as environment, cultural patterns, historical determinism, statistical norms, or the hundred and one vagaries of the pseudo-science of psychology, the modern white magic of evasion.

Praise is equally difficult to apportion. We learn that our best actions are the result of the basest motives, and our basest ones are prompted by factors beyond the individual's control, and are, therefore, not essentially culpable.

We resign ourselves to the necessity of living with Uncertainty for the rest of our existences. In this limbo, effort is the result of an over-functioning gland, and apathy the natural outcome of sluggish enzymes. Praise is either a corporate or a national expediency, and blame is what Johnny saw in the cellar at the tender age of three.

The whole box of tricks is like predestination without the courage of conviction, original sin without moral fortitude.

In his work, the waterman is delivered from the rat's maze of uncertainty. The villains and the heroes of his story are concise and unambiguous; the issues are cleanly cut. Either he has the courage to endure, or he has not. Either he can apply accumulated knowledge intelligently in a trying task calling for will power and the desire not to be defeated, or he can not. He knows his villains well enough. Wind, sea, tide, and weather alternately defeat him and are conquered by him, and both praise and blame are equally easy to apportion. When he wins, he feels a stronger sense of satisfaction than most of us ever have occasion to know. The basic lines of his contest are clean, clear, and uncluttered.

Perhaps this fundamental satisfaction of the waterman's job has been most clearly expressed by Albert Klestadt in his book entitled *The Sea Was Kind*. He is in the middle of escaping alone in a twenty-foot sailboat from the Philippines under Japanese invasion. He writes:

> Now, for the first time in my life, I felt completely at peace with myself. . . . I saw myself engaged in a clean and clear-cut fight. I was my own master, and salvation depended only on me. If my seamanship . . . my strength and wit were good enough, I could get through. This voyage was a test of strength, and there was great satisfaction in tackling it.

Interest

One often reads of the essential monotony of most of the tasks performed in a highly industrialized society. I remember watching a man at work in a bottling plant. He sat on a white stool in a white jacket, and his own face was also white, as if recently sterilized. He sat quite still and stared endlessly at a line of filled and capped bottles. It may be that he breathed, but only the bottles moved, precisely, without end, so many a minute, every minute of every hour.

To a man accustomed to using his full strength and most of his faculties in a variety of tasks, a job like that must seem, in Norbert Wiener's phrase, "an inhuman use of human beings."

The waterman's jobs are as various as the seasons. He tongs in winter, eels or fishes nets in spring (sometimes both), crabs in the summer or takes out fishing parties. The same man who digs clams in the summer very often removes his clamming ring and tongs in winter or sets nets in the spring.

What escapes the outsider is the almost passionate interest felt by watermen in their various trades, and especially in that of tonging. Everything to do with oysters, from tonging to marketing, is, to them, a subject of the most vivid and eternal freshness, even after fifty years on the water.

If a group of tongers happen to be working fairly close to one another on a bar, the chief subject of their conversation can be safely predicted. After a good deal of mutual chaffing, most of it satirically critical, yesterday's gossip is handled briefly: Joe Higgin's father-in-law died last night in convulsions after an enormous supper of fat back and greens. His pathological symptoms are then examined in some detail. But someone mentions that a mutual acquaintance fell overboard yesterday while oystering on The Lumps, the same one who went over the week before. It is the consensus of opinion that, while he can save himself bath-time that way, he won't catch oysters thrashing around in the water. The subject is now oysters and oystering, the price, the new set, the old set, the number of young oysters present on Hambleton Hills eight years ago, the tidal difficulties of tonging The Ravens, the relative effectiveness of straight heads as opposed to gooseneck heads, long rakes versus short ones, the price of oysters per bushel at Broadcreek, on the Tred Avon, the Honga River, at Crisfield, and so on for the rest of the day. They are genuinely and always interested in every possible aspect of their work, near or remote.

Furthermore, the tonger is absolutely determined to catch as many oysters as he can wherever he can find them; and he will search them out with almost fanatical determination. When the fleet comes in to sell, and a tonger finds that he is a bushel or two under the average catch, he says frankly that he has "made a failure"; and he means it. Competition among experts creates its own excitement. Men who are consistently good at their work are widely admired and respected.

There are grades among watermen as there are in other professions. Those who are less competent, but work hard and willingly with the best, are liked and accepted. The slacker, the shiftless tonger with no will and courage, is tolerated with dryly humorous and good-natured contempt.

If you happen to have a bushel of oysters and you want to keep them, don't put them overboard; eat them or give them away. The tonger, in his determination to find oysters, figures that everything on the bottom is his.

In 1953, the Chesapeake Biological Laboratory at Solomon's Island instituted a series of oyster growth experiments on trays placed on the bottom. When the trays were examined for growth of oysters in 1954, it was found that no losses from wave action had occurred, but that all oysters over three inches in length had been removed from the trays. Now, in Maryland, the legal length for salable oysters is three inches. The Institute report comments rather dryly:

It is doubtful that satisfactory results can be obtained from such growth studies until a suitable area is found that can be protected from both marauders and storms.

Well, yes.

Some oyster bars are so badly crowded because of large numbers concentrated in a single area that the competition among oysters for space and food is very severe. The result is that the shells of these oysters are misshapen, thus making them difficult to shuck, and the meats are thin and of poor quality. The market value of such stock is low.

As an illustration of the general concern among oystermen for every facet of their trade, it is interesting to note that in 1955 the tongers of Queen Anne's County, on a voluntary basis and without pay, tonged and transplanted twelve thousand bushels of overcrowded oysters to better growing grounds. These oysters were taken from Prospect and Crab Alley Bays and placed where more room would result in better quality.

The interest in bivalves spreads from the water to the land. Wherever he goes ashore, the tonger is stopped by every other acquaintance and answers searching questions as to the relative abundance and quality of oysters. Since, as I have said, oysters produce, annually, more food and more income than any other commercial product of the Bay, merchants and businessmen are not entirely indifferent to the subject.

Even the public school system has more than an academic concern for tonging and dredging. One-third of the tonging license and one-half of the county dredging license receipts are retained by the counties to be used for local schools.

In 1952, the Deal Island High School founded a club, called "Future Watermen," and acquired five acres of ground under the 1951 law which authorizes the State to set aside a submerged tract for each public high school for the cultivation of oysters. The boys in the club planted shells to catch spat (young oysters) and got a harvest of seed oysters. Some of the young oysters were left on the bottom and grew to market size for harvest in 1955. They were doubtless harvested, but whether by future or by present watermen is not known.

Skill and Knowledge

While a vivid interest in their work is real and general among watermen, it should be said also that their jobs are in themselves anything but monotonous and mechanical. If a man thinks that all he needs to be a successful oysterman is a strong back, he would do better to stay ashore. Knowledge, and much detailed knowledge, is essential. A tonger has literally to "learn" many hundreds of acres of ground in widely separated localities of quite different kinds, ground he seldom actually sees. Furthermore, what he knows one year may be of limited use the next, because of changed conditions on the bars. An oyster bar is an organic entity, subject to the fluctuations common to living organisms. On the other hand, what the tonger learned eight or ten years ago may now produce high dividends, for he has learned that it pays to remember absolutely everything he has ever known. But he must also be intelligent. He must be able to adapt immediately to change and apply his knowledge under new circumstances.

It should also be taken into account that conditions of work are seldom the same two days running. Areas which can be worked profitably during a very low tide are time-wasters at the top of the flood. Grounds which are very rewarding in a northeast wind are often virtually worthless in a northwest breeze. Every change of weather and fluctuation of tides will raise different problems which call for different solutions.

These are not matters which are mastered within a week or two, no, nor in a year or two either. No matter how good your muscle tone is, if you have little knowledge, or if you are unable to apply what you know, you will never make a professional waterman.

I was forty years old before I ever picked up a pair of rakes. In my first week as a tonger, I made precisely four dollars. I didn't know enough. I had a general idea where to look for oysters, but general ideas are not enough. When, by luck, I did drop the rake heads on a little patch, I was unable to catch them. I raked them up into a pile, all right, but when I lifted the shafts, no more than two or three oysters came up stuck in the teeth and nothing in the heads at all. I didn't know how to close the rakes properly on the bottom. In fact, I didn't know anything. After eight years of tonging, I still have much to learn.

For one thing, I am still rather shaky in the skill of getting marks. By that, I mean that when a tonger finds a nice little patch way off nowhere, he does not holler, and whoop, and jump up and down. He keeps his mouth shut and he works like blazes. Before he quits for the day, he takes his marks at every cardinal point of the compass. He lines up a farm house on shore with a big pine tree, a silo with a barn, a break in Clash's Woods with a water tower, Herring Island buoy with another on Hambledon Hills. When he comes back the next day, he will find his patch.

I saw a man lose a fifteen-pound grapnel one day on New Bar. Since he had no other anchor with him and the afternoon was well on anyhow, he started his motor and left for port. The next day, he cruised carefully on the edge of New Bar, watching his marks. He dropped his new hook and made fast. Within ten minutes, he had tonged up his old anchor.

It sounds easy. Try it sometime over miles of water.

On Saturday mornings, we generally work only half a day, and it is a matter of some pride to bring in a decent jag for a morning's work. On one Saturday, the tide was exceptionally low, but on the turn, with a strong flood running. Considering the conditions, I calculated that the best place to work upriver would be the edge of Pine Bar. When I sold, I put out three bushels. Captain Johnny Schnaitman, seventy-one years of age and fifty of them on the water, put out five bushels. He had remembered that the inside bar of the horseshoe at Second Point is a good place to work under the conditions I have described. So Captain Johnny out-thought and out-worked me.

Another order of skills is needed to work on stone beds. These large collections of rock occur on some bars and often occupy most of their area. They are not composed of a single kind of rock, as might be thought, but of heaps piled helter-skelter and consisting of limestone, sandstone, shale, quartz, and feldspar at random. One encounters occasionally even serpentine and other igneous materials. Since I have never found any specific geological account dealing with their origin on our bars, I am at liberty to guess that they may represent detritus carried in the past by ice. The stones range in size from small pebbles to rocks as big as a dining table.

At any rate, these acres of stone often prove attractive to spat and are highly productive of oysters, so it pays to work them. As you can imagine, these beds are difficult to work, since the teeth of the rakes constantly snag in stones.

A young fellow whom we will call Jerry was in his first season of tonging. He had purchased and equipped a small boat, and one Saturday morning, having heard rumors of oysters being taken on Bennett's Point, he decided to go down the river and try his luck. Several boats were already there. Since he had never worked the bar before, he anchored near the little cluster of workboats and right on top of stones.

Jerry put the rakes over, and something like a spasm of disbelief passed over his face. I had a box seat, for I was not fifty feet away from him at the time. When he felt that rough, knobby ground, he must have thought he had dropped his heads on a patch of big oysters. He began to rake. And he raked. After a while, he lifted his load and he could barely get it out of water. There was a sound like a rumble of thunder when he dumped the lick on the board. Rocks bounced everywhere.

Jerry stood over the culling board while the rake shafts slipped through nerveless fingers until the heads rested on the bottom. He looked around. Everyone in the cluster of boats regarded him silently.

Jerry said, "You fellows on stones?"

Everyone nodded. One man said, "If it's stones you want, this here is the right place."

Jerry waved an arm in a circle. He said, "Everywhere?"

"Not everywhere, no. Maybe twenty, thirty acres. Should be enough."

Jerry was a game kid. He watched while a man next to him pulled up a whole rakeful of clear oysters; and then he went back to work, and he worked hard. On one occasion, he got his rakes up halfway and had to sit down on the cabin top and get his breath before he pulled them the rest of the way up. He had a rock as big as his head with one small oyster attached to it. Everywhere in the boats men bent double over their rakes hoping they wouldn't bust, not on a Saturday morning and a weekend coming up. I didn't hear as much as a chuckle; a few gasps of pure pain, but not so much as a smile.

They let him make his mistakes, and he made them all in rapid succession. In an hour's time, he had a board full of rocks and twelve oysters. The men liked his pluck, but of course they never told him so. That would never do. Instead, Jerry began to get a hint here, and one there.

"Don't go to rakin' and rakin' a big pile of stones up on the bottom. You won't lift 'em nohow."

"Just take a little lick, and bring her up."

"You got too much tooth on them heads. You want stobs for this work."

"Feel for them. You feel oysters, take a short lick. Ain't no use heavin' rocks up all day."

By the end of the morning, Jerry was catching oysters. When he has worked rocks two or three seasons, he will make a good stone bed tonger.

1 AUTHOR STOOPING TO LIFT RAKES IN FOREGROUND CULLING BOARD WITH THREE-INCH NOTCHES USED TO MEASURE OYSTERS

2 TONGER WITH A RAKEFUL

3 RAKES JUST EMPTIED

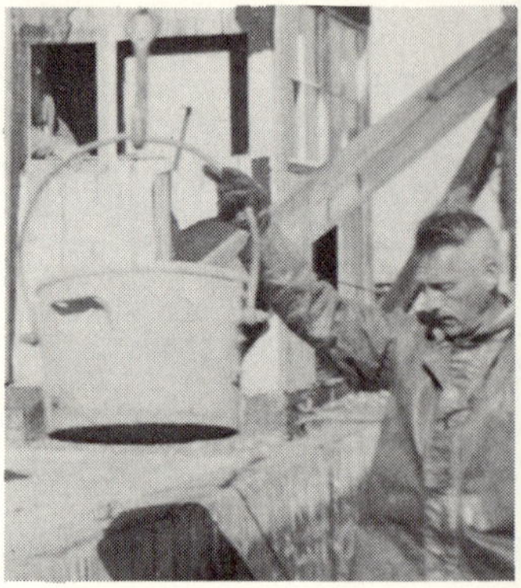

4 IRON MEASURE USED BY PACKERS AND DEALERS

5 TONGERS SELLING TO TRUCK

The fineness of touch which allows a man to distinguish between stones and stone-like oysters is developed gradually through practice. It is an interesting example of a faculty which has made man successful in the use of tools, and it is called tactual extension. We extend the nerve ends of our fingers and hands through the inert material of the tool and actually "feel" with it. Tonging is an extreme instance of tactual extension. Through twenty feet of wooden shafts and a steel rake at the end of it, oystermen extend the nerves of their hands and "feel" the ground. In this way, they are able to make extraordinarily delicate distinctions among materials encountered by the rake teeth. An experienced tonger can often tell when his rakes are resting on shell only or on oysters growing on shells. In this case the materials are precisely the same, but the "feel" is slightly different. How to describe so fine a distinction in "feel" is difficult. When asked, a tonger may say that the "feel" of oysters is more "spongy"; the rakes have a little more "bounce." But these are highly inaccurate distinctions compared to the touch itself.

When the season first opens in September, and for the whole of that month, all a man needs who just wants to pick up a few oysters for eating is an ebb tide, and the tide obligingly ebbs at least once during every day. You find a place where a few scattered oysters occur on the flats, and you wade for them. You just stoop over and pick up oysters as a farmer plucks ripe tomatoes from the plant. I have often gathered a bushel in that fashion in the course of a couple of hours. It is necessary to know where to look for them; otherwise it is quite possible to wade for miles without finding any.

Later on in the season, of course, wading is a chilly business, but the frost comes to the aid of the sighter. Cold precipitates much of the floating matter in the water, and it is often possible to see bottom in anything from six to ten feet of water. In the shoal water of some bars, there are enough scattered oysters at that depth to make it commercially worth the time and effort to catch them, and they are usually taken by nippers. These consist of two shafts, crossed like a pair of scissors and pinned together like tongs. Unlike tongs, the two shafts do not come entirely together when the nippers are closed, for what is wanted here is a hard grip for lifting single oysters. Nippers are armed with a pair of paws. The word "paws" is vividly descriptive of their appearance and function. If a bear had four claws on one forepaw and three on the other, and if he fitted them neatly together to pull fish from the water, his paws would closely resemble those on a pair of nippers. The shafts are as long as fourteen feet or as short as eight.

The problem now is to look down through six or eight feet of water and "sight" for oysters. As in other aspects of oystering, much practice is needed to become adept at nippering, for oysters are frequently well-disguised.

One day in September, while wading for crabs, I said to a friend of mine, "Paul, just pick up a few oysters and put them in your dipnet, will you? There aren't any crabs in here; I'm going to look out on the edge of the flats."

When I came back half an hour later, Paul was still wading and looking. There was not a single oyster in his net, and he was walking on them. To an unpracticed eye, they look for all the world like small, irregular stones, roughly oval-shaped. They feel like stones to the feet. Even after I had picked one up and shown it to him, Paul was for some time unconvinced. To one not used to the sight, oysters piled on a culling board resemble a heap of bottom trash.

The nipperman, after choosing a place where he can see bottom, shoves his boat slowly over the ground and peers down, hoping to see it littered with fine, big fellows, waiting to be seized by the paws poised at the end of the nippers. He may see nothing. He may have passed over a hundred oysters without having seen a single one, so cleverly are they camouflaged as small rocks, clumps of algae, clusters of mussels, bunches of barnacles, and bits of sponge.

But a practiced nipperman can do better, when the water is clear and the air calm, than a tonger can. I have known a first-class nipperman to catch as high as fifteen or twenty bushels in a day's time.

The best way to learn the skills of oystering, and I have by no means covered them all, is the slow and early way. The young fellow fifteen or sixteen years of age who goes out with his father or an older relative will be a competent tonger by the time he casts his first vote. Nothing will ever replace in effectiveness the apprentice method when it comes to acquiring the skills of a waterman. The consequence is that watermen seem to run in families, and when you see a Haddaway or a Harrison, a Hall or a Higgins, you will know that his blood has been before him on these waters, and that his skills are inherited.

conservation

2

In the Skin

A Talbot man, let us call him Ernie, once decided to visit some relations in western North Carolina. Since Ernie was a first-rate tonger and generous by nature, he decided to take along with him a bushel of oysters as a present. He anticipated the pleasure of the recipients of such a fine gift during the long drive from St. Michaels to North Carolina. But he was much disappointed when he got there.

His relations, far from being overjoyed with the present of a bushel of prime oysters, ignored the basket for some time. Finally, one of them turned to Ernie, nudged the bushel of oysters with one foot, and inquired, "Ernie, do you b'ile them things in the skin?"

The Animal

The oyster is a much more complex organism than most of us suppose. When we open the shell, we see little but a lump of greyish white matter, having about the consistency of a well-set jelly. And yet, this lump of jelly has gills, stomach, kidney, esophagus, heart, and muscles. He is very nearly as well-equipped as you who eat him. The Bay species also has two sexes.

In the summer months, when the temperature of the water has risen to seventy-five degrees or thereabout, eggs are ejected by the millions and they are fertilized by the male sperm. The result of this conjunction is a free-swimming larva which eventually, if not eaten, which is generally the case, attaches to some hard, smooth object, such as a stone or another shell, and begins at once to grow a shell of its own and to go about its business of being an oyster. The business of being an oyster consists of securing as much food as possible by opening the lip of the shell and generating a current of water through the gills by means of a rapid oscillation of tiny, hair-like fibers on the gills. In this way, gallons of water are passed through the gills in a day's time, while the gills themselves select minute food particles and pass them on to the hungry mouth. In about three years' time, the spat or young oyster will be of legal market size: three inches from hinge to lip.

For the major part of every day, oysters keep their lips open; that is, the two shells are slightly parted. If you look down in clear water, it is possible to see them in this position, dozens of oysters with lips parted, all feeding hungrily. If you lower a stick gently and give such a feeding animal the slightest touch, he closes his shell at once, for he is sensitive to touch and vibration.

Anyone who has tried to shuck an oyster knows how powerful the round disk of muscles in the middle of the shell can be. An acquaintance of mine shoved his skiff out to his boat by pushing an oar blade against the bottom. He felt something heavy attached to the blade and drew it up to look. At the end of the blade a large oyster clung, his lips closed so tightly over the wood that it was difficult to knock him free. That oyster had been caught completely unaware while feeding.

Oysters tend to congregate on parts of the bottom which are best for propagation and growth. The accumulations of shell and oysters are called bars, and it is over them that the tonger anchors and plies his rakes. For the natural formation of a bar, at least two requirements must be met: there must be hard, comparatively smooth material to which the spats can attach, and there must be food. The first requirement is generally supplied by stones and the accumulated shells of the generations of oysters which have died, to say nothing of those still living. This entire mass of matter on the bar bottom is called cultch or "rock."

I have found as many as a dozen spat on a single large oyster shell. The spat will attach to glass, pebbles, or rocks, but shell is by far the most attractive. I have seen a pair of abandoned iron rails for launching and hauling up boats literally plastered with oysters, but it is significant that no shell in any quantity was nearby. The spats had no choice. If given one, they almost invariably prefer shell.

The second requirement, food, is supplied by minute animals and plants: protozoa, the spores of algae, and the tiny plants called diatoms, all of these so small as to be scarcely visible to the naked eye. Rain, which washes fertile chemicals into the water, adds materially to the nutriment upon which these tiny plants and animals feed. They multiply freely, and the oysters fatten. So both farmers and watermen like rain in the summer time.

The Enemies

Like almost every organism on earth, the oyster is a prey to other life forms. One of the most devastating is the oyster drill, a small marine snail that bores through the shells of young oysters and thus destroys them. Fortunately, in Talbot County the water is not sufficiently saline to support the drill, but it has done enormous damage in the lower part of the Bay.

For example, in the Chincoteague Bay area where an oyster known all over the nation for its quality and flavor has flourished for generations, very grave depredations have occurred due to the oyster drill. From a production of 500,000 bushels in 1910, the catch fell to 50,000 in 1951, a devastating loss which continues annually. In some instances, private planters in the area have noted the staggering destruction of up to eighty and ninety per cent of the oysters planted to replenish the failing beds.

But a hitherto unknown virus disease which has, in the last two years, virtually wiped out rich beds in the Delaware Bay, has now struck Chincoteague and threatens to complete the annihilation already in process due to the oyster drill. The Chincoteague oyster may soon become a name only.

During the season of 1959-60 and again in the fall of 1960, Queen Anne's tongers working out of Wye River were being paid as high as seven dollars a bushel for prime oysters in baskets. They sold their catch to a truck on the banks of the Wye. It was an unheard-of price, and Talbot tongers watched the Queen Anne's men filling baskets with a feeling of pure envy.

In the fall of 1960, the trade in basket oysters suddenly stopped. I learned later that these oysters were being shipped by truck to Chincoteague Bay, where they were put overboard for a week or two and then gathered for sale as Chincoteague oysters. The law stopped this practice, but it shows to what desperate expedients the planters in that area have been driven.

In addition to the drill and the virus, a fungus (Dermocystidium) occurs. This parasite is not harmful to man but is a well-known killer of oysters. It is widespread in the Virginia waters of the Chesapeake. Samples of Maryland oysters were taken throughout the Bay in the fall of 1954, and the fungus was found as far up as Solomon's Island, and in the Sounds of Tangier and Pokomoke. High salinities and high water temperature both favor the growth of the fungus. Again, Talbot County is fortunate in not having waters sufficiently salt to support this disease.

Water pollution is not as severe a problem in the Chesapeake as it could be, for water fouling has been, in other areas, a major cause of oyster destruction. The loss of the Long Island Sound bars and the famous Blue Point oyster is a case in point.

In the thirties of this century, Long Island was pitted with duck ponds, and every restaurant, tavern, hotel, and hamburger joint advertised and sold Long Island Duckling. It was a booming business. There was a house-cleaning problem, of course. With millions of ducklings in hundreds of ponds, an enormous quantity of feces collected. These were washed or flushed into the Sound by way of the ponds. So the famous Blue Points were killed or polluted by duck dirt.

Storms sometimes cause limited damage on oyster bars. After hurricane Hazel plowed up the Bay and its tributaries in 1954, the Department of Tidewater Fisheries made a survey of possible damage. It was found that private planters suffered most, since they tend to grow oysters in shoal water. In shallow areas, especially on the Western side of the Bay, many oysters were buried or washed ashore. In comparison to private beds, most of the public or natural bars appeared relatively undamaged.

But Chincoteague again took a beating. There ninety-mile-an-hour winds churned up the soft bottom and carried oysters as far as one-half mile from their beds.

Considering the severe natural hazards to which oysters are subject, it seems only rational that men should take some measures to conserve them. But before men become generally aware of a need to save, they must first destroy.

Time

The earliest tool used by professional oystermen for the taking of oysters was the same as that used today by modern tongers. Obviously, a tool which is two hundred years old is an antique and therefore inefficient. But the plain fact of the matter is that tonging is much easier on natural beds than is dredging, for example. It is claimed, and with considerable weight of evidence, that a hundred years of hand tonging alone have not depleted a single natural bed, whereas dredging has destroyed some of the finest bars in the Bay.

Oystering is the one trade in the country in which efficiency is to be shunned and deplored. On the whole, the less efficiency in the tools for taking oysters, the better for all concerned, oystermen, packers, and consumers. A glance at the history of oystering in the Chesapeake will, I think, prove my point.

The early depletion of the New England oyster beds affords an interesting example of wholesale destruction. From the very beginning of their foundation, New England colonies used oysters, not only as food, but as raw material to make lime for fertilizers and building purposes. In an amazingly short time, once fertile beds were so badly depleted that the colonies concerned passed a number of laws restricting the use of oysters to food purposes only and prohibiting their export. These conservation measures were far too late and too little. Shells from shucked oysters continued to be used for other needs than bar rehabilitation. By the year 1800, many of the biggest and best New England beds were completely exhausted.

It was then that Yankee skippers heard persistent rumors of vast and practically virgin oyster beds in the lower Chesapeake Bay. A wide-awake master of a schooner coasted from New England to the Chesapeake and there offered to buy all the oysters local tongers could catch. It was a very lucrative cargo he took back home with him. By 1808, many Yankee vessels were carrying Virginia oysters to New England.

But Yankees who could get some always looked for a way to get more. Tonging was too slow a process to suit them. They brought with them to the Chesapeake the same dredge which had helped to ruin their native bars, and they began to catch oysters for themselves. This method of taking oysters was then quite unknown in the Chesapeake. Virginia and Maryland oystermen, once they saw the speed and efficiency of the new technique, proved just as greedy as the Yankees and soon adopted the dredge as a regular means of harvesting oysters.

The State of Virginia felt the teeth of the new dredge first. It faced a dilemma. It was known that the inefficient tongs made smaller catches but did not get *all* of the oysters on a bar by any means. While overjoyed at the increase of oyster catches, the State was much disturbed over the fact that the dredge's efficient blade caught every oyster in its path, leaving no mature stock for future replenishment of the bars. Virginia passed a law prohibiting the use of the dredge.

Maryland was bitten next. The Virginia law pushed the New Englanders into the upper Bay, and Maryland was forced to prohibit all dredging in 1820. Due to pressure put by dealers and tongers on the legislature, Maryland then legalized the use of the dredge in the waters of Somerset County in 1854, and in Dorchester and Talbot counties by 1874. Dredge boats are still employed in these counties on limited areas set aside for their use.

The next step of the Maryland legislature, and a fatal one, was to allow unlimited use of the dredge in all State waters of the Bay proper over fifteen feet in depth. The depth limit was almost impossible to enforce, and dredge raiding of tonging bars became a common practice. The same bill excluded the use of steam as a motive power. But what the sail dredge had done in New England could also be accomplished in the Chesapeake.

In one delirious decade, from 1880 to 1890, the Bay and its tributaries produced the gigantic harvest of twelve million bushels a year. Workboats were literally filled to the washboards twice a day, and the bugeye dredge boats could scarcely swim under the weight of cargo. Jubilant watermen saw a golden future and an inexhaustible supply. They wanted no interference from the law. The packers and seafood marketmen agreed.

But a dramatic reversal of fortunes brought with it a change of view. After a peak twelve million bushels in 1890, production fell to three million and then to one. The great harvest was over, and once magnificent beds were almost totally depleted in the rapacious way that men have. But a lesson was learned.

The two chief reasons for depletion were unlimited dredging and a reluctance on the part of merchants and packers to return empty shells to the bottom. The uncontrolled use of the dredge plus waste of the shell is more than any bar, however rich, can possibly support.

Gradually, tentative steps were taken to preserve what was left of Maryland's oyster industry. Dredging was limited and supervised. A cull law was introduced which required oystermen to cull over the bottom where they caught their oysters. It was forbidden for oystermen to keep or transport either shell or oysters under the size established by law, at first two and one-half inches; later, three. In this way small oysters and shell were returned to the bottom where they belong.

But, in the vital matter of bar rehabilitation by returning empty shells left over from shucking to the grounds from which they were taken, little was done. The State slumbered for a long time, woke up briefly to make an arrangement for purchasing from packers a mere ten per cent of the shell left after shucking, and then nodded off again. Still, this law helped to check at least some of the terrible waste.

I have seen shell heaps over two stories high shipped off to be burned in lime kilns.

The United States has such a dramatic history of colossal waste in the matter of natural resources that it is famous the world over among resource management men, ecologists, and conservationists, who point to one horrible example after another with something approaching awe. But even in a country renowned for mass spilth and spoliation, Maryland's misuse of shell must remain fairly conspicuous.

In the 1920's and the early thirties of this century, almost every county road in such areas as Talbot, Kent, Queen Anne's, Dorchester, and Somerset was made of oyster shell and nothing but oyster shell. These dazzling white roads criss-crossed the tidewater counties for some thousands of miles, and visitors thought them "quaint" and "typical." They were typical, all right, but the word "quaint" never occurred to those who had to live with them. In dry weather, the light, pulverized dust was such as you have seldom witnessed. Trees, bushes, every blade of grass, the very cattle in the fields, for a hundred feet either side of the road, were thickly coated with shell meal. Wet weather resulted in a thick, greasy paste.

The ditches were normally half-filled with oyster shells flipped from the crown of the road by the wheels of every passing car, truck, or tractor. By this means, road flooding was positively assured after every heavy rain. I have known Tunis Mills to be cut off for three days running when the shell of the road was once more back in its native element. Without a thorough weekly

6 A TYPICAL OYSTER OR PACKING HOUSE

7 DREDGERS AT WORK HAULING IN THE DREDGE

dragging by a team of county mules hauling a V-sled (later by a tractor and blade), these roads would have been virtually impassable in the best of weather.

So millions upon millions of bushels of shell were wasted in the construction of inferior roads, simply because the material was cheap and readily available. What we did not grind to meal on the roads was thrown into dry wells for domestic sewage disposal. We did our best. The fact that oyster bars survived at all is a great tribute to their natural richness and productivity.

It is a pleasure to report, then, that the State has at last wakened completely to the absolutely vital necessity for preserving and planting shell. As the result of a law passed by the General Assembly in 1953, the State now has a claim to fifty per cent of the shells left over after shucking in each oyster-packing plant in the State. The consequence is that shell planting has increased substantially. Some of the more enlightened packers let the State plant all of their shells, instead of the legally required fifty per cent. A great stride has been made in the right direction. Our county roads are now either made of sand and gravel, or they are hardtop, an improvement in every respect; and the shell pit for sewage is now illegal.

Shell Planting

Of course it is known that shells planted in some areas do not necessarily receive a high set of spat the first year. However, when plantings are made in waters where high sets are frequent, a set of spat will eventually occur and ultimately produce oysters having a value far in excess of the cost of planting.

These shell plantings are made by large power barges hired by the State. Shells are loaded from oyster house middens in late June of every summer, just before the first period of oyster spawning. The large, wide decks are piled high with shell and transported to the ground selected for planting. There they are washed from the decks with a powerful stream of water, which makes for efficient scattering and saves the time required for shoveling by hand.

The official view of the State of Maryland as presented by the State Board of Natural Resources in 1958 is as follows:

> The production of oysters in Maryland is dependent largely upon natural replacement of oysters through the attachment of spat on existing bottom cultch.

Obviously, where cultch has been entirely depleted, no set can occur, for there is no shell upon which the larval form of the young oyster can attach itself. If cultch is meager, it follows that setting of new oysters is likewise curtailed.

In order to estimate the amount of set which occurs each summer in every major oyster-producing area of the Bay, the Department of Tidewater Fisheries makes a count. It does this job by two means: first, by taking samples from the various grounds; and, second, by placing wire baskets of shell in certain stations throughout the Bay. These baskets are pulled up and examined for set in the fall of the year.

Annual spat counts throughout Maryland show a wide variation. A few localities have consistently high rates of successful reproduction, and low sets in these areas occur rarely. In this respect, as in that of the natural enemies of oysters, Talbot County is very fortunate. During the twelve-year period 1942-53, the average set per bushel of shell was 138 in the Miles River and 307 for Eastern Bay. These were the highest average sets for any area of Maryland. The set on the Western Shore of the Bay averaged for the same twelve-year period 100 spat per bushel in the most productive area and zero in the least.

Again in 1957, Eastern Bay and the Miles River showed a set far above average for the entire Bay. Other areas likely to receive good sets annually are the Choptank and Potomac Rivers, Hooper Strait, and Tangier Sound.

Some areas, such as those on the Western Shore, seldom have successful reproduction and the rate of spat attachment is quite limited. Let us say that a particular bar sets consistently only 10 or 15 spat per bushel of cultch on the bar. If all the spat survive, the result in three years' time will be 10 to 15 market oysters per bushel of cultch. But it takes approximately 400 three-inch oysters to fill a bushel. A bushel of cultch, then, has produced only a very small fraction of its potential. The result is an annual loss, in terms of the cultch present on the bar. Additional shell planting on such a bar is largely wasted.

Tongers in such Bay counties as Talbot, Dorchester, and Queen Anne's like a rainy summer. They know frequent rains will fatten the oysters, thus producing prime stock at prime prices. Ironically, the very rains that fatten our oysters bring death to thousands of bushels on the bars in the Upper Bay and in the uppermost bars of the Bay tributaries. In these places salinity is lowered to the point where the oysters can no longer live. A further irony lies in the fact that it is precisely in such areas of low salinity that cultch is most abundant. During periods of decreased rainfall, the necessary environmental conditions do occur, and these areas often become highly productive. They get enormous sets. Then the cycle is repeated. Heavy rains and runoff from melting snows pour freshets from watersheds, and oyster mortality is severe. But the enormous amount of shell formed by living oysters during the dry part of the cycle remains, millions of small tombstones, so to speak.

On the other hand, those areas having good yearly sets of spat are the very ones where the supply of cultch is either completely absent or sadly inadequate. The supply is rarely found in sufficient quantity, and it continues to diminish. Large areas of many Talbot bars are as bare as a table top.

Cultch losses result from many causes, such as siltation and heavy seas which tend to bury shell or scatter it. A shell-destroying sponge, *Cliona celata*, reduces shells to small fragments unsuitable for spat attachment. But the chief cause of loss remains in the yearly harvest.

One answer to the problem of oyster conservation is immediately obvious. In order to *maintain* an average annual oyster production, it is not only necessary to return *all* the shell back to the bottom, but to place it on those areas where it is likely to produce a reasonably good yield.

In order to *increase* production annually, *more* shell must be planted each year than was removed from the grounds.

So far, it has been a struggle to maintain, let alone increase, production. We are still paying dearly for the oyster orgy of 1880-1890 and for our own tardy and inadequate measures since then. The total catch from Maryland waters in the season of 1957-58 was 2,190,074 bushels, the lowest since 1950-51; and of that number of bushels caught, only 1,454,194 were returned to the grounds.

If legislation is passed requiring all shucked shell to be returned to the State, instead of fifty per cent as at present, the prospect of maintaining annual yield is good. To increase the amount of shell planted over the quantity annually removed is another problem. An important step in that direction has recently been taken by the Tidewater Fisheries Commission, which has located and is now dredging an old shell bed in the lower Bay. This bed is buried and non-productive, so that all of the shell taken from it and planted on fertile ground is pure gain. In addition, there is also the possibility of removing cultch from areas subject to frequent mass mortality due to rainy cycles and planting this shell in areas which are annually productive.

Lease and Public Ground

I should say at once that wherever I have written of tongers in the previous pages, I have meant those men who work with their own gear in their own time on public bars, not on privately owned or leased grounds. Unless the tonger himself owns such leased ground and works it himself, he is not the free waterman I have so far described: he is as much a hired hand as any field or factory worker. He does what he is told to do when he is told to do it. He is not a free tonger.

Actually, as far as quantity goes, the problem of distinction is not difficult. In 1958, there were 273,800 acres of public oyster ground in the State, as opposed to only 9,915 acres of privately leased grounds. Public grounds can be tonged by any citizen of the State of Maryland who has the knowledge, the skill, and the license to work them, within the county where he lives and purchases his license.

The production of oysters on public or State bars is so far in excess of that produced by private planters that the latter are scarcely worthy of mention in terms of total fertility.

Nevertheless, the subject of conservation would not be completely treated without a consideration of claims advanced by private planters to the effect that, acre for acre, they produce more than the public bars. It should be said at once that such claims have never been satisfactorily substantiated. In fact, the history of private grounds in the State seems to show that they suffer as much as, if not more than, public grounds and for the same causes. You will remember, for example, what has happened to the largest single holding of private grounds in the State, Chincoteague Bay. Nor is it for lack of either shell or seed oysters that private grounds have experienced severe difficulties in maintaining annual production, let alone stemming a progressive loss of fertility. Every year, the State of Maryland sells thousands of bushels of seed oysters to private planters. In 1952-53, the State sold over twenty thousand bushels of seed to private planters, without any significant rise in production in the years following. In 1955, over thirty-one thousand bushels of seed oysters were sold to private planters, with a loss of total production in the Bay for the following years.

The fact is that private grounds are subject to the very

same fortunes as those of contiguous public grounds, and for the same reasons. Where the drill is bad in a given area, it is bad everywhere. The fact that a name is placed on a stake over private grounds is no guarantee that snails can read. The only possible way to increase production of oysters in the Bay is to increase production on natural bars; that is, in those areas which have always been, because of favorable environmental conditions, good producers for a hundred years or more, time out of memory. A natural bar does not happen by chance. It occurs because conditions favor its development. And these are the very areas which are now both public ground and the best producers.

I have nothing against the leasing of grounds to private planters, if those planters work their own leases. I do strongly object to leasing ground to men who sublease to packing houses or other corporate or company enterprises. New grounds should remain either entirely public or in the hands of watermen who will work and develop them on an individual basis. In fact, I think the State should encourage watermen to lease small areas of barren ground wherever suitable areas occur. This practice might very well aid in maintaining annual production. Public bars, however, should remain inviolate. They constitute the last major holdings of highly fertile public lands worked in common by the free citizens of the Free State of Maryland. This is a splendid tradition which should not be carelessly and willfully destroyed.

The Law

No discussion of conservation would be complete without some mention of the law enforcement branch of the Tidewater Fisheries in Maryland. Its history, like that of oystering, is an old one. The present Department of Tidewater Fisheries is the direct descendant of the State Oyster Police, created in 1868. To this day tongers, when feeling a little less than courteous, refer to the enforcement officers as "oyster cops." They are known officially, however, as the Marine Patrol and Inspection Force.

Most of these officers have always been as good watermen as any afloat; in fact, their ranks have been filled for generations from the families of watermen. When they began their work, they were in sail like everyone else. In 1894, a newly-built bugeye was launched under the name of *Brown Smith Jones*. This ridiculous nomenclature resulted simply because the Governor of Maryland at that time was named Brown; the Comptroller, Smith; and the Treasurer, Jones. It is a great wonder she didn't founder on launching. She survived these names, however, to become part of the State Navy, and was used to aid in the enforcement of oyster laws. On her deck forward of the foremast, she carried a one-pound, quick-firing rifle.

I tell you, they took law enforcement seriously in those days. In these days, the role of enforcement has been recently belittled by the present Chairman of the Tidewater Fisheries Commission, Mr. H. C. Byrd. Says Mr. Byrd:

> We've been spending too much money on law enforcement that doesn't enforce, and not enough on development.

Certainly, more money spent on the development of Bay resources would be welcome. But no conservation is possible, let alone development, without a strong enforcement agency. For example, with weak law enforcement, the temptation to catch and sell oysters that are under the legal size is almost insurmountable. Heedless and dishonest tongers would soon force the large majority of steady and law-abiding men into the same malpractices of which they are guilty. The first consequence of weak enforcement would be a serious shortage of future breeding stock left on the bars on the one hand; and, on the other, a depletion of those oysters which, if allowed to remain on the bottom, will grow sufficiently during the summer months to become next season's harvest. The results of feeble enforcement, in short, would be disastrous. Instead of development, progressive deterioration would be a certainty.

In addition, the old evils of over-harvesting, even now a grave danger on private grounds, would soon be felt on public ones as well, if present laws against the use of scrapes and dredges were not energetically enforced.

Mr. Byrd also implies that the Tidewater Police do not enforce the law. What is obvious to packers and watermen alike has evidently escaped Mr. Byrd: and that is that all of Maryland's efforts to maintain an average annual production of oysters of from two to two and a half million bushels would have been totally wasted without the efficient enforcement of the oyster laws on the part of the Tidewater Police. The tremendous effort these men have made for years, and their excellent devotion to duty, deserve the highest praise, rather than censure or belittlement. In fact, if I had set out deliberately to undermine the morale and the efficiency of the Tidewater Police, I would make just such a statement as that of Mr. Byrd.

In 1956, the law enforcement staff consisted of ninety-seven men. Their equipment was forty-one patrol boats, a half-dozen outboard motor boats, and one seaplane. With these meager means, ninety-seven men are expected to patrol and enforce all of the fisheries laws along 2,000 miles of shoreline and on 2,437 square miles of tidal waters. This is a man-sized job. That it is done at all, is amazing. That it is performed with vigor and resolution, is altogether admirable.

Aside from apprehending and prosecuting violators, the Tidewater Police carry out many other services. They take water samples for the Health Department to determine degree of pollution in tidal waters. They assist the Coast Guard in locating missing boats and people and aid in the recovery of drowned persons. They also aid the Department of Game and Inland Fish by making counts of wild waterfowl, and they direct the State's shell and seed plantings. Finally, they are expected to furnish information on any and every subject to the general public.

All things considered, it may not be surprising to learn that these men are often on duty seven days a week. What may seem a trifle uncommon in a country addicted to the forty-hour week is the fact that they receive no extra pay for extra hours.

The enforcement officers are well grounded in the elements of their jobs. Each summer, they undergo a week of intensive training in such subjects as Legal Evidence, Courtroom Procedures, Legal Aspects of Conservation, Fish Identification, Life-Saving, and First Aid.

The St. Michaels watermen have been very fortunate in having had for many years a succession of excellent officers. They watch us like hawks. That is their duty. But they also are ready and cheerfully willing at any time to help any waterman in trouble; and the nature of his work is such that trouble is not unusual. An injured man, a man whose boat motor will not run, can always expect help from other watermen; but if there are none near him, he knows that the oyster patrol will come by at least once a day, and that help will be at hand.

crabbing

3

The Animal

A crab is like an oyster in only one respect: he wears his skeleton on the outside. If you add that both make excellent eating, you have then run out of comparisons. Their growth, for example, is quite different, for where the oyster prefers to grow quietly, gradually, and all of a piece, the active crab has only one way of getting bigger, and that is by abrupt, successive stages: explosions of growth. The crab molts or sheds its entire outer shell or skeleton, including carapace, both top and bottom, legs, and backfins. After shedding, he is one-third again as big as he was before. He simply bursts out of his horn-like, protective covering. He sheds approximately once a month until full grown, after which shedding no longer takes place. I have caught mature crabs whose shells were covered with moss and barnacles. The shells of younger individuals are comparatively clean, since the old shell is soon discarded for a new one.

As soon as the crab has cast his shell, and for several

hours thereafter, he is a soft-shelled crab, since hardening of the new shell takes time. He can then be eaten shell and all, and is known as a soft crab. A crab that has just begun to molt, one whose shell has split at the back and shows the soft skin of the new covering beneath, is known as a "buster," a word which perfectly describes his condition. But there are even earlier indications of a coming molt, shown by the joints in the legs and backfin (appendage for swimming) and by a minute cracking in the suture of the lower carapace. Crabs in this stage are known as "peelers," and are, as you know, often sold for fish bait. They are unexcelled for attracting rock, hardhead, and other marine fishes. The flesh holds together on the hook, since the fine, soft membrane of the new shell has already begun to form beneath the carapace.

Busters and soft crabs do not feed. But crabs about to shed are hungry. They need to stoke up for a period of fasting and inactivity. The consequence is that crabs just about to "peel," or "green peelers," are frequently caught by professional crabbers. These crabs are perfectly good cooked in the usual ways, or they can be sold for bait.

Crabs that are shedding generally come into shoal water and conceal themselves in sea grass. Their greatest enemies when they are newly shed or soft are, in this order, hard crabs, striped bass, and man. A male hard crab that stumbles upon a concealed male soft crab has found food. A male hard crab that has run across the hiding place of a female soft crab has found love. A crab's psyche must be amazingly uncluttered.

The best way to take soft crabs and busters is to wade through shoal water on a bottom well grown with sea grass. Once you see a soft crab, it can be caught easily with a dipnet, for it is not as active at this time as when hard. Freshly caught, soft crabs are excellent food, meaty and delicious, though I, for one, prefer the hard crab.

I have taken sides in many a controversy as to the time of month when soft crabs are most plentiful and, therefore, most easily caught. There are two schools of thought among us, which can be roughly described as the rational and the astrological. The rationalists point out that, in order to grow, a crab must shed: that he will shed any time his shell can no longer contain him. To support their argument, they invite attention to the fact that soft crabs have been caught every day of every summer month, tide in and tide out.

The astrologicians, while admitting that soft crabs occur at random, claim that they are at the peak of their abundance at or near the full of the moon, on the few days before, during, and after. Close to thirty years of wading for crabs have convinced me that the astrological school is correct, though I will not go so far as to say positively that the full moon is a direct cause, as it is said to be in the shedding of human inhibitions.

It is well known, however, that spring tides accompany a full moon, thus assuring for several days of every month an adequate depth of water on the grassy flats which crabs favor for shedding. High water allows the molting crab to move in close to shore and away from enemies. After a full moon, I have often found dozens of fresh sheds or cast shells in the space of a few square yards of bottom. I believe also, again from observation, that the crab tends to shed at high water. I will go further, even at the danger of angry contradiction. I believe that the favorite daily hour for shedding is in the morning.

If I and the astrological school are right, we can now fix optimum times for the soft crabber, whether professional or amateur: 1. At or near full moon. 2. Just after the top of the tide. 3. Preferably after high tide in the morning.

You understand, of course, that after having fixed upon the perfect day, even the hour, of the month, and prepared dipnet and boat with confident expectation, you will catch no soft crabs unless hard crabs are plentiful; nor will you catch any if crabs are present and a northeaster at the same time. It is a gamble any way you look at it.

Just for the fun of it and to get something good to eat, hard crabs can be caught in a number of ways, but my favorite is to wade for them with a dipnet. With this method, you are in the crab's element, and you have to be quicker and more cunning than he is. A hard crab can move like a flash. If he has been dipped after before, he is wild and wily, and he will give you some sport. You have seen comic postcards from Ocean City showing a big crab gripping a big toe in his claws. Unless you step on a hard crab in water with your bare foot and keep your foot on him, that is never known to happen. A crab in water will slash at you with his claws if he is run down or cornered, but he will not grip and hold. He is too busy attempting to escape. He will hold like the jaws of a steel trap out of water, but not otherwise. Nevertheless, it is a good thing to wear an old pair of shoes when wading because shells can split your foot open as neatly as a razor blade, and small pebbles leave a stone bruise you will not forget in a week's time.

Most Shoremen prefer crabbing from a boat with a long-handled dipnet. They claim what is perfectly true, that the additional height affords better visibility. Mobility is acquired by shoving the boat with the handle of the net. I think my preference for wading stems largely from the fact that I am the least-skilled skiff crabber the Shore has ever seen. I have will, enthusiasm, and good eyesight. All I lack is ability. That I have will and enthusiasm, at least, is proved by the fact that I have more than once, during the heat of the chase, hurled myself overboard. I always pretend that I meant to do so, but that is not strictly true. Sometimes, the boat and I part company without warning or mutual agreement.

My neighbor, Les Daffin, is the most accomplished skiff and dipnet crabber I have ever seen, and I have looked at some good ones. I have often watched him shove quietly around the shore of an evening and bring back half a bushel of big hard crabs and half a dozen soft ones. Les can look down and see a hard crab in grass where I can not see one, even after he has pointed him out to me: the tip of a backfin or the point of a claw, a line of slowly rising bubbles, these are all the hints he needs.

The Great Feast

During every season when crabs are plentiful, from late May to late October, from Somerset to Kent, the entire Eastern Shore enjoys a long and glorious feast. As soon as hard crabs appear, men go out after work in the evenings, or on Saturday afternoons and Sunday mornings, and they come back with hard crabs, enough for them, their families, and their friends: the great summer-long feast has begun.

Free food, gentlemen, and what food! These are not of the trifling kind whose claws only can be eaten. Every shred of the Maryland blue crab, claws, legs, backfins, and body, can be enjoyed, with the exception of the shell (when hard) and the lungs (dead man's fingers). The fat is prized as an additional flavoring. A bushel of seven- or eight-inch crabs (measured across the top shell from point to point) means literally pounds of white, flaky meat whose flavor and texture are second to nothing that lives in the water, with the possible exception of the terrapin, the snapping turtle, and the lobster, and even with these there is room for argument.

Steamed hard crabs are eaten right out of the shell, as you would eat peanuts, one leading automatically to another. Some families seldom bother to eat anything else all day Sunday and every Sunday in a season when crabs are plentiful. These are partly social feasts, for friends are always willing to pick a crab with you, but largely they are communal, a part of the eating habits and the diet of an entire area.

However, purely social occasions on the Eastern Shore tend to function around the steaming and consumption of hard crabs. If you want to see a crowd of happy people, go to a crab feast on the Shore, where crabs are served freshly caught and freshly cooked, as they generally are not in restaurants. There are dozens of tables, all spread with newspaper; a whole mountain of steamed hard crabs heaped high in Olympian abundance; and a clatter of wooden hammers on claws, a rousing, rattling din that precludes all attempt at talking. In any case, there is no time for talk. As soon as one crab is picked and eaten, another waits. The empty, pillaged shells pile up and are removed in wash tubs, while another heap of crabs is poured in careless abandon on the table top. Quart after quart of beer provides relish and the cool liquid necessary to withstand the heat of red pepper and Old Bay seasoning. The word "feast" describes such an occasion as no other could. Many a man has staggered from the crab tables, glassy-eyed, swag-bellied, and supremely content.

Professional Crabbing

After he hangs up his rakes on the fifteenth of April, the tonger may decide to work ashore in a boat yard for a while, or he may begin at the beginning of crabbing. There are no edible crabs available. They are still sound asleep in the mud where they buried late last fall. What the hard crabber has to do now is to catch horseshoe crabs instead.

In warm weather, these crabs are easy to find. If you look down at the sand in shoal water, you will see long, irregular tracks, about the width of an ordinary pencil. All you have to do is to follow one of these tracks and you will find the fellow who made them by dragging a long, horny, spiked tail. His name is descriptive, since the large forepart is exactly a horseshoe in shape, the abdomen being small and flat. Some of them are a foot or more across the shell of the forepart.

The spiked tail looks wicked but is never employed for attack. It is a kind of trailer coupling. During the breeding season, the male hitches on to the spiked tail of the female, and sometimes another male will hitch on to his tail, and all three will move slowly over the sand, although what the second male either hopes for or receives is beyond me. In fact, I have no sound understanding of the horseshoe crab. He is certainly a kind of crab, but he is also just as certainly a kind of spider, of which a glance at his eight spidery feet would convince anyone. Very crawly indeed. Only, some of his feet are not feet at all, but small pincers or claws.

Although the horseshoe crab lives always under water, he is an air breather. Just how that is managed, I could not say, but there it is. He is said to be absolutely inedible, and he is also said to be good eating. I, for one, have no desire to eat a kind of spidery crab that breathes air under water, although hogs eat them with great relish and a luscious crunching of leathery shells and multiple feet.

To crown the horseshoe crab's ambiguity, he should not be here at all. He is a living fossil who stubbornly refuses to recognize the fact, a very backward and primitive life form indeed, halfway between one species and another.

Some thirty-five million years ago, these life forms or ones like them covered the floors of shallow Jurassic seas. They might have become the precursor of the true crab, but the horseshoes never got there. Neither did they become what they might have been with equal logic, a scorpion. They just got hung up somewhere in between and have stayed there ever since.

The Bait

Horseshoe crabs are not going anywhere now, except into a crabber's drag. In April, the water is still cold, and the horseshoes are out deep on the mud, so dragging for them is necessary. When the crabber has gathered about a barrel of horseshoes, he drops his grapnel and gets to work. He has left his culling board on the boat, and he reaches into the barrel, picks up a horseshoe by the tail, slams it down on the culling board, and chops it up with a hatchet. Since the females are packed with eggs the size of bird shot, this job is likely to be a little messy. Eggs fly everywhere and stick wherever they strike. For a hatchet man, boots, skins, and sou'wester are recommended apparel. Of course, if you are squeamish, you can go alongside the nearest maninose dredge and buy a bushel of clams, but that costs money.

Having chopped a big pile of bait, the next thing to do is to lift the hook and run along the shores of whatever cove, creek, or head of river is likely to have a large eel population. Having selected the site, the crabber slows his boat to walking speed and baits an eel pot with horseshoe crab fragments. He throws the pot over in shoal water, and the cork buoy attached by cotton line to the pot marks its position. By the time he has the next one baited, the boat has moved sixty or seventy feet, and he tosses over another pot, and so on, until all of his pots are baited and over. He will run anything from fifty to two hundred pots.

The pot is made by watermen of galvanized wire of one-third- to one-fourth-inch mesh in the form of a cylinder. It has two funnels forward and a sleeve at the rear which is tied to close the pot. The bait is placed just behind the first funnel. The eels go into the first funnel for the bait, and then through the second funnel to the rear of the pot.

After the pots have fished a couple of days, the crabber collects and chops more bait, and then fishes his pots.

Again he slows his boat to a walk, grabs the cork buoy of the first pot, hauls it in by the line, opens the sleeve at the rear, dumps out the trapped eels into a barrel, and rebaits the pot. By that time the second cork is floating at the side of the boat. He throws the rebaited pot over and picks up the cork of the second pot, all in one motion. And so on, until the entire string has been fished. If he has done well, he may have fifty or sixty pounds of "green eels."

His day's work is by no means ended. He now has the job of preparing the eels so they will stay sweet through the hot summer months. He pours a quantity of salt from a hundred-pound bag on top of the eels and adds water. This mixture of salt, water, and writhing, twisting, leaping eels he stirs with a paddle until he has a heavy brine. The brine accomplishes three things: first, it kills the eels; second, it sloughs off the outer membrane of the fishes, which makes them so notoriously slippery; and, third, it begins the process of salting. After letting the eels soak in brine for several hours, the crabber next washes them quickly in fresh water to complete the cleansing process begun by the brine. He then rolls or tosses the fish in salt until they are well covered and places them in a barrel for storage. The salt gradually works into the flesh of the eels, thus preserving them for later use. I know all this sounds a bit like Betty Anne's Television Recipes for Better Angel Cake, but I don't suppose any housewife in the country would want to be up to her elbows in eel slime.

At any rate, the eels are now clean and dry, without a trace of slime on them, sweet and easily handled, their skins of the consistency and texture of very fine leather. These tough, fine skins can be beautifully tanned. In the eighteenth century, they were tanned and worked to make elegant sheaths to contain the tails of gentlemen's perukes or periwigs.

A salted eel is virtually inedible. The very fish that we catch, salt, and throw carelessly into barrels by the ton every spring are, unsalted, one of the greatest of delicacies for the table in Europe. However, if our population continues to increase at its present rate, we are going to learn different eating habits, and eels with their firm, white flesh may re-enter the American cook book.

But the crabber has no idea of eating eels. His object in catching them is solely to get bait for crabs, for he knows that a hard crab, like Europeans, loves to eat eels. The bait is perfect for its purpose. It is easily handled and sufficiently tough and firm to hold well on a line and last under the sharp little choppers of the hard crab.

Most watermen plan to get enough eels, say two or three barrels, to see them through the crabbing season. But others are specialists. They trap eels for a living, their object being to catch a surplus to be sold on demand. Since some crabbers always buy their bait, and many others run out of it before the season ends, there is almost always a demand for salted eels. Your professional eeler starts to trap earlier than the waterman who simply wishes to meet his own needs. And so, while I was still tonging last spring, Captain Maurice of St. Michaels was already eeling. I was working on Pine Bar that spring, and Captain Maurice used to pass me on his way toward the upper Miles. About ten-thirty or eleven o'clock in the morning, he came by again on his way to port after fishing his pots. And every morning as he passed he held up one finger, meaning "one barrel of eels." That summer Captain Maurice sold three tons of eels; but he "sold too close." Before the season was over, he had to put out pots again to supply his own need for bait.

The Trot Line

Along in the middle of May or as late as the middle of June, depending on how they move, crabs are in sufficient numbers to make commercial fishing profitable. Now all that work that was done in the spring begins to pay off. The professional crabbers bait their lines with chopped pieces of eel and get to work.

The line itself is slender, stouter than fishing cord and smaller than clothes line. Like clothes line, it is made of cotton. The baits are tied directly into the line by putting slip knots in the cord itself about a fathom apart. At the end of the line, a piece of chain is attached in order to sink the baited cord and keep it on the bottom, for a crab is a bottom feeder. To the chain end another piece of line about twenty-five feet in length is attached, and that leads from the chain to a buoy. From the buoy another twenty-five-foot length leads to an anchor, generally an old piece of iron. The other end of the trot line is equipped in the same manner.

When the crabber lays out his line, he drops the anchor first, then the buoy, and then the chain. He proceeds to feed out the rest of the baited line until he comes to the chain at the other end, which he lets go overboard, followed by the buoy and then the anchor. The only things showing above the surface, then, are the two buoys, one at either end of the line. The line itself may range from one-quarter of a mile to a mile in length, and the distance between the two buoys will correspond to the length of the line.

After the line is laid, the crabber fools and fusses around in his boat for a while to give the line a chance to fish. He arranges his baskets, gets his dipnet ready, and puts a roller out over the side of the boat. He approaches the buoy with his boat slowed to a walk, dips up the line leading from the buoy to the chain holding the line on the bottom, and places it on the roller. The line slips over the roller and back into the water as the boat advances, bringing up the chain and after that the first bait on the line. When the line is raised over the roller in the moving boat, one bait after another comes to the surface as the boat moves forward. The crabber dips the crabs under the surface of the water as they rise with the baits from the bottom.

Crabs hold on to the bait and come up best in the very early hours of the morning. The consequence is that the crabber lays his line in the pitch dark before dawn. The most exasperating things can happen, even to an experienced waterman, due to almost total lack of visibility. Consider, for instance, the shattering experience of one St. Michaels man. He started to lay out a mile of line in the river in the black of late night. He threw over the anchor, and the buoy, and the chain; only the chain and line snubbed in his rudder and caught there. As he advanced cautiously through the thick dark, feeding out line, he was dragging the light anchor, the buoy, and of course the chain behind him. When he got to the end of the line, he dropped off his buoy and anchor as usual, then pulled back on his steering stick in the middle of the boat to sheer away. Pulling back on the stick shifted the rudder, naturally. When he shifted the rudder position, he jarred loose the line that was snubbed there all the while. This waterman saw something completely out of nature and all common sense when the light of dawn allowed him to start to work. As he picked up his line at the buoy, he stared straight ahead, reason tottering on her throne and the imagination boggled. Not fifty feet away was his other buoy; and he had laid out a mile of line.

One expects this order of experience in a sweat-wringing nightmare, not in the sober, workaday world. It is a great tribute to this waterman's essential stability that he is still with us, still going calmly about his various tasks upon the water.

The dark can play any number of tricks, all of them infuriating. If you lay out in close quarters, such as a creek or a cove, you have to be a good judge of distances. You have to have a feel for it, since darkness distorts the normal sighting and estimate of measurements over water, difficult enough under the best of conditions. You may find that after you have fed out all but fifty yards of line, you are no more than fifty feet from shore. You then have two choices: you can take the entire line up and re-lay it; or you can cut the line. Since valuable time is lost in re-laying the line, you generally use the knife. That means the chain, buoy, and anchor have to be retied on the end of the cut line.

One source of bitterness and violent argument among watermen has been largely removed by law. Crabs often tend to concentrate in particular areas. When that occurs, many watermen are drawn to the same grounds at the same time, and the resulting tangle of lines can be an unholy mess. If you have gotten up at two in the morning to be first on the ground, then you are likely to be very angry if another crabber comes along and lays his line across yours. When you fish or "overhaul" your line, you are lifting the weight of his, thus making it impossible to crab the whole middle section of your lay. You are now so hopping mad that the temptation to cut the overlaying line with a knife at the point where it crosses yours is irresistible. Your competitor, when he makes his first overhaul, notices that his line is so slack, there being no anchor to hold it in place, that he, in turn, is unable to work. And then the fun begins. And not so funny, either. I have known men to carry shotguns in their cabins for just such occasions.

The law now reads that the first line down has priority, and that it is illegal to cross it with another line. It is a very wise provision.

The steering stick I have mentioned is placed in the middle of the boat near the washboard and just aft of the roller. It is a necessity for crabbing. It is obviously impossible to keep running back and forth from the roller where the crabs come up to the tiller in the stern of the boat. So lines are made fast to the tiller and run to a stick, by means of which the crabber can steer the boat along the line with one hand while the other holds his dipnet ready.

If you have not fouled your line in heavy sea grass, if you are not running your line too fast or too slow, if your line is neither too slack nor too taut, if you have not kept the sun behind you and so frightened the crabs with your shadow, if your bait is fresh and sweet, and if there are any crabs to catch, you will make a living of sorts. One thing about it, you will always have crabs to eat.

At about nine-thirty or ten o'clock of a bright summer morning, you take up your line and head to port with your catch. While you are running to market, you continue to tap your steering stick occasionally with one hand while you rebait your line. Every bait is quickly checked as you feed your line into a barrel where it is stored until fed out again the next morning. If baits have been badly eaten by the crabs, they are replaced by new pieces. Any baits that show signs of softness or deterioration are also tossed away and replaced. As your boat moves toward market, a constant stream of old or half-eaten bait goes overboard, and you and your boat advance in a cloud of screaming, wheeling, fighting gulls, all eager for the discarded eel. It is good fun to throw an old bait high in the air and watch a dozen laughing gulls rush for it. One of them in full flight picks it from the air with his bill, a feat of dainty, flashing precision.

While crabbing, you have already sorted your catch into jumbos and culls. The jumbos are put into bushel baskets with tops over them to contain the crabs, which, of course, are still alive and kicking. These crabs are from seven to nine inches across the carapace from point to point, and they bring the highest price. The smaller crabs or culls, from five to seven inches, bring very little as a rule; so little, in fact, that I steamed, picked, and froze the meat of the culls I caught last summer. We now have crab cakes in January.

The jumbos are loaded onto trucks and taken that same morning to Baltimore, Philadelphia, or other large outlets.

After selling, you moor your boat and complete the baiting of your line. You go home, eat your dinner, and then try to take a nap in the afternoon. You have been up since three in the morning, and tomorrow you will get up at three again, so an afternoon nap is highly advisable. Most watermen find sleeping in the afternoon very difficult. They tend to get far too little sleep in the course of a summer. Many of them lose weight on what is otherwise a comparatively easy job, barring bad weather.

Pot Crabbers

The crab pot is used for commercial crab fishing in the Bay or State waters. It is illegal in county waters. These pots are made completely of chicken wire, have four funneled openings at the bottom, a bait container in the middle, and a hinged lid for emptying the captured crabs. Any lure can be used, but the commercial bait is the much-despised alewife, a trash fish bought frozen by watermen. The pots are dropped in anything from fifteen to sixty feet of water, oyster bars being excellent ground. A light line with a buoy attached marks the position of the pot for future fishing.

Sometimes amateur crabbers who haven't the skill to take crabs any other way will make illegal use of crab pots in county waters. These pots are often "lost," permanently. Trot-liners hate and despise them. The pot you set one evening with the height of hungry hopes may very well be gone the next day.

The Law

Almost every State in the Union has found it necessary to protect its natural resources by law, and Maryland is no exception. For many years, people from Baltimore, Philadelphia, Wilmington, Washington, and other large urban centers came in impressive numbers to the Shore for the sole purpose of laying out miles of trot line in order to catch crabs. The increased use of the boat-trailer finally brought this practice to a point of organized week-end plundering on a very large scale. Competition with commercial crabbers in the various Eastern Shore counties became so severe that, in 1955, the Maryland General Assembly enacted a measure requiring a license of all crabbers using a trot line of more than one hundred yards in length. These licenses require a sworn and duly notarized statement to the effect that the licensee is a resident of the county in which his license has been issued, and the license is good for that county only.

In seasons of ordinary abundance, one hundred yards of trot line will give plenty of leeway for fun and food as well. George Mullikin, an enforcement officer based at St. Michaels, devised a means of rapidly measuring unlicensed trot lines. He mounted an ordinary fishing reel on a sawed-off rod. On the reel he wound strong, light wire exactly one hundred yards in length. The end of the line was provided with a snap hook. So all Captain George has to do is to snap the line to one buoy and run the inspection boat toward the other one, letting the line reel out as the boat moves forward. If the unlicensed line is over one hundred yards in length, it is confiscated, and the user pays a fine.

In other respects the law on crabs is simplicity itself. It fixes the legal size of hard crabs at five inches; of soft crabs, at three and one-half inches; and of peelers at three inches. "Sponge crabs," those females bearing eggs in a spongy ball on the under carapace, are not legally caught.

The Crab Catch

In 1957, from April 1 to November 30, 28,375,013 pounds of crabs were taken commercially in Maryland. This figure does not include those taken for food instead of for market. Keep in mind also that this measurement is not of crab meat, but of pounds of crabs in the shell.

Of this amount, 5,474,379 pounds were taken in Talbot County alone, a figure exceeded by only two other counties: Dorchester, with 11,058,800 pounds; and Somerset, with 7,616,422 pounds.

The men accounting for the total Maryland catch were 786 pot crabbers and 2,539 trot-liners, or a total of 3,325 commercial crabbers. This handful of men really caught a heap of crabs.

Of these men, 1,440 worked on the Western Shore and 1,885 on the Eastern Shore of the Bay. Considering that 1,885 Eastern Shore crabbers caught almost twenty-seven million pounds of crabs, while 1,440 Western Shore crabbers caught only a million and a half, I think it is fair to say that we whipped them. They apparently had fisherman's luck over there: a wet tail and an empty gut.

Conservation

The chief problem in conservation is caused by the fact that no average annual catch can be depended upon. The fluctuations are very marked, and may appear from one season to the next. In 1953, for example, crabs were so plentiful that by the middle of the season packers had to restrict the catch by closing the market two days a week. Most sections of the Bay were literally alive with crabs under market size. All indications were for a bumper harvest in 1954. But in that year, production fell from twenty-six million pounds to less than twenty million, the lowest catch since 1948. In 1955 the catch was the lowest on record since 1944, a mere fifteen million pounds. But by 1957 the catch was up to twenty-eight million. In 1959, however, the Bay saw the worst crabbing season in its history, positively disastrous. Crabs in almost all areas did not occur in commercial numbers until the end of the season. In the middle of that season, I saw trot-liners overhaul a mile of line without ever catching a crab, and many a man had three hard crabs (crabs, not bushels) to show for a morning's work. Salted eels were worthless and tons of them were thrown away as garbage at the end of the season. In 1960, however, crabs were back again in commercial numbers, though they were not plentiful.

The official view of these wild fluctuations in hard crab production is contained in the 1957 report of the Department of Research and Education:

> Present evidence indicates that little effective control of these fluctuations is now possible, since natural factors, and not man's fishing efforts, appear to be the major cause of fluctuation. The hatch and survival in some years are much greater than in others, and it is not yet possible to isolate the controlling conditions.

In other words, these fluctuations are an act of God. But Maryland watermen are perfectly certain that the Devil lives in Virginia and there does a dance of death every winter, for it is in this state at the mouth and in the lower part of the Bay that female crabs bury in the winter season. Virginia allows winter dredging of buried or hibernating crabs. A letter from Mr. John W. Bates, President of the Tred Avon Waterman's Association, printed in the *Easton Star-Democrat* for May 27, 1960, gives a neat summary of the problem:

> As we are told, the crab originates at the mouth of the Bay. As they begin the trek up the Bay (in the spring) as small crabs, the obstacle they encounter are

> large schools of striped bass. . . . If any crabs escape and enter the tributaries, they then become the prey of man. If they still survive, they will become adult crabs.
>
> As an adult, the female crab will start to migrate in the fall for the mouth of the Bay to bear young in the spring. As she leaves the tributaries and enters the Bay, she is lucky to escape the crab pots. If she can still make the trip to the mouth of the Bay, she will settle in the soft mud.
>
> When the water becomes cold, the crab scrape goes into operation. And by spring there are very few females left to bear young.

While oystering, I have, on occasion, inadvertently tonged up male hibernating crabs. It is generally recognized that to disturb the normal hibernating habits of any animal is to put it to grave hardship which it may never survive. A crab disturbed in hibernation is quite helpless. Its metabolism has been so lowered that it is incapable of fending for itself, and the temperature of winter water alone is enough to kill disturbed, unburied animals. Exposed and helpless, it stands almost no chance of survival. I think it is reasonable to suppose that for every crab winter-dredged, there must be fifty that are exposed and disturbed, though not actually caught, and left helpless on the bottom with little chance of survival.

While watermen know that male crabs bury at random, they are certain that females hibernate in Virginia. On this subject, the Maryland Department of Research and Education will speak only with the most tremendous reserve:

> It is known that impregnated female crabs move South in the late summer and early fall.

That is all that the Department can be made to admit. But whether Maryland watermen are right or wrong about Virginia, we have a little house cleaning of our own to do.

Peeler crabs are often sold as bait, as I have already mentioned. Another practice, though, is to hold them in floats and wait until they shed, because soft crabs bring a higher price than hard crabs. The trouble here is that crabs shedding in an unnatural environment often die, especially if much disturbed. The loss of the upper shell, for example, before shedding is complete, is generally fatal. In addition, some of the crabs are not so far advanced toward molting as others, or they are slower. A peeler, especially a "green" one, will attack and totally consume a soft crab, which, as you can see, has no way of protecting himself by hiding in the confined space of a float.

Unless wire of fine mesh is used to enclose the float, a practice by no means common, swarms of minnows can and do consume soft crabs in an amazingly short time. Before I knew any better, I once put a buster down in a peach basket to allow him to continue to shed. He would not, needless to say, shed out of water. By that evening, nothing was left of my crab but his empty carapace, proving he had shed during the day, and a little bundle of resistant integuments and fibers.

It is impossible to keep busters, or any crab, for that matter, in stagnant water. Many a man has splashed water into a skiff to keep a few hard crabs overnight, only to find them all dead by morning. So the float must be used. And even with the float, certain conditions of tide and calm can cause stagnation and consequent mortality. From time to time, huge kills take place in floats and are duly reported to the Chesapeake Bay Institute for Biological Research, and these losses are investigated by trained personnel.

In late July of 1953, to take a single example, heavy mortalities were reported in crab-shedding floats located in the Fishing Bay–Tangier Sound area. One company alone reported the loss of 30,000 peeler crabs overnight. The personnel of the Institute found that very hot weather, light winds, and an extremely low tide all were prevalent for several days in the area reporting the kill. The floats, normally staked in three to five feet of water, had ten to twelve inches of water under them at the time of the kill. Heat, lack of water circulation, and decrease in the amount of water present, had caused a severe shortage of oxygen. The crabs were asphyxiated.

Occasionally, crab potters, especially those who work the deeper waters, report losses. These are also attributed to low oxygen content of the depths in which the crabs are trapped and must stay until fished by the crabbers.

There is no need to mince matters. The truth is that there is a great deal of waste on the water. Much of it is unavoidable, part of the normal hazards of harvesting, but every bit of waste which can be stopped should be stopped; otherwise our resources will gradually dwindle, and finally disappear.

clamming

4

The Animal

The clam considered in the following account is not the hardshell clam or quahog. He is a soft-shelled animal whose outer covering is so brittle it can easily be broken by pressure from your fingers. It is called in the Bay area the maninose, a corruption of nannynose, presumably so called because the clam has a nose or siphon.

Years ago, we often used to dig them from the sand of shallow water in order to get fish bait. If you wade in the shallows on a calm day, you can look down and see the bottom pitted with irregular holes about the size of a pencil. Insert a forefinger in the hole and see if you can follow it down. If you can, lurking at the bottom of the conduit is a live maninose. It is through the hole that he obtains a circulation of water necessary for food and oxygen. By means of the pipe, you will track him to his bed. Of course, you have to dig, usually through a layer of fine gravel, and you are digging with your hands. From eight to twelve inches below the surface of the sand, your fingers will encounter a soft but rubbery object which is the nose or snout of the clam. Dig carefully along the sides of both valves to disengage the animal whole and uncrushed from impacted sand. Lift him against the suction formed by the water and silt of the narrow hole occupied by the plunger of a fist filled with maninose. Bring him finally and in triumph to the surface. You have accomplished something. You have gotten a single maninose. As you see, it is not easy.

After a day or two of turning over sand and fine gravel for maninoses, a curious sensation develops in the hands, especially the fingertips. Quite unnoticed, a slow but very efficient abrasive has rubbed the skin to tissue thinness. The nerve ends, exposed and irritated by constant erosion, now respond to every touch with hyperacuity. The world, once visual and aural, is now tactual. To lift a piece of paper is to experience an exquisite tingling akin to both pleasure and pain; but the rough surface of a linen table cloth is anguish. The very thought of manila hemp is a cause for shuddering. You may talk about the tortures of salt pickles and mumps, but you ought to try well-sharpened fingers and emery cloth.

On a spring tide at ebb, it was possible at one time to take a shovel and turn up clams with it, but their shells were usually crushed in the process. As I say, they were dug for bait, not for eating purposes. Oh, there were a few people who ate them, yes; but they were thought of as having a perverse appetite. They were pointed out as you would speak of a man who habitually eats snakes, in a sort of revolted whisper:

"You see that man? He eats maninoses!"

The same prejudice is operative today. Our clams are eaten, not by Shoremen, but by New Englanders.

Ways and Means

In the forties of this century, it was finally discovered that New Englanders would not only eat clams; these peculiar people would actually pay good money for them. It was an astounding piece of news and started the soft-shell clam industry in Maryland.

At first, the means of digging clams commercially was very primitive indeed, just a little better than going after them bare-handed. They were gotten by an ordinary workboat in no way different from the ones employed by crabbers and tongers. But the boat was worked in a different way. The clammer set up stout, tarred nets on the sand flats in three or four feet of water and then secured these nets by stakes driven into the sand. He then anchored his boat so that the stern pointed directly at the middle of the net, about eight or ten yards away from it. Having securely anchored his boat aft, both port and starboard, he started his boat engine. As the boat propeller or wheel turned, it agitated the water, which scoured the sand and maninoses from the bottom. The clams were driven by the wheel stream into the net, at the bottom of which they fell and piled up.

The clammer worked the territory carefully in front of the net by shifting either his anchors or his anchor lines. When the area was thoroughly worked, the motor was shut off, and the clammer went overboard, protected from the water by body boots. He pulled a skiff or rowboat to the net. There he used an ordinary dipnet to pick up the clams thrown against the net. The maninoses were placed in the skiff, and later sorted and packed into baskets.

Outboard motors placed on movable frames were also used, but the principle is the same: hydraulic mining of clams.

In work of this kind, there were many disadvantages, but I will consider only two.

The first is that a good deal of wading about in body boots had to be done. If a sea were running, a slight bending of the legs to get a purchase on the bottom or a slight increase in depth was sufficient to send a wave splashing into the boots, which were difficult enough to work in without having them swamped with water. On windy days, clammers had to work in the lee, and even then a boat's wake could cause them trouble.

The rubber armor they wore extended from the toes of the feet to the middle of the chest and higher. It was inexpressibly oppressive in summer, cold and clammy in winter. And yet, the clammer had no choice. Skin exposed daily and for long hours to water soon becomes wrinkled, porous, and of little use as a protective covering. Watermen have a very expressive word for such water-logged skin: rotten. And of course exposure of skin to water in winter is out of the question. So the clammer wore his armor.

The second disadvantage, this time from the standpoint of the clam, is more serious. It must be obvious at once that this type of hydraulic mining cast large quantities of sand over quite a large area surrounding the focus of operation. In order to live, the maninose has to have a pipe through which to draw water, and from which to expel it. If several inches of sand are piled over his pipe, he smothers. Tens of thousands of clams were destroyed in this way.

It is not likely that anyone will ever again see this rather curious means of making a living, and it is perhaps just as well.

The Hydraulic Dredge

In New England, because of a greater tidal range, the soft-shell clam is dug from large, exposed flats. The difficulty in Maryland was that the tide normally does not range above a foot either above or below the median. The result is that few tidal flats of any appreciable area are ever exposed. Add to that the obvious difficulties of taking clams from underwater beds by the methods I have described, and you have the chief reasons for the fact that the clamming industry remained stagnant.

In 1950, however, an Eastern Shore waterman by the name of Fletcher Hanks developed and later patented a hydraulic dredge that harvests clams efficiently in depths of water up to eight feet. The dredge consists primarily of an endless chain-link conveyor belt attached to a sled-like structure called a scoop. The end of the scoop is lowered to the bottom, and a pump on board the boat directs powerful jets of water backward toward the blade of the scoop and the conveyor belt. As the boat and blade move slowly forward, clams are mined out by the jets and sent backward to the belt, which, in turn, conveys the clams upward to the side of the boat, where they are picked up by the operator of the rig and placed in bushel baskets.

Now, the clam dredge requires a considerable investment of money to begin with, nor is it cheap to operate. It has three motors which are in constant operation during the actual dredging: one to run the boat, one to move the endless belt, and one to pump water for the jets. But it is an extremely efficient clam harvester, and the result of its development was a tremendous boom in the clam industry.

The catch of clams in 1947 amounted to 6,200 pounds of meat. By 1955, the catch had risen to 1,293,720 pounds of meat, or over two hundred times as much, and ninety-one clamming rigs were digging trenches twelve to eighteen inches deep wherever their scoops could touch bottom. They also produced, in addition to large clam harvests, one of the bitterest feuds among watermen the middle area of the Bay has ever witnessed. Only the calmness and good judgment of the marine patrol officers and the native restraint and common sense of the watermen themselves prevented open violence.

The Feud

Apparently men are seldom able to exploit a new natural resource with new equipment and not leave a trail of wanton destruction in the area of operation. One reason for the havoc thus wrought is the lack of legal restraint in the beginning of a new industry. From 1950 to 1955, the clam dredge and its operation was almost completely uncontrolled. As in the sudden development of any new natural resource, money was being made rapidly and with comparative ease, and neither clam-buyers nor dredgers wanted any interference from the law. If this sounds familiar, you are right, and I refer you to the history of oystering in the Bay.

Only one major legal restraint was put upon clammers: they were not allowed to dredge on oyster bars. But the State, with its tendency to sleep upon problems rather than to deal with them, was caught napping again, and this time for a period of fifty years. In other words, the official charts showing the location and extent of the public bars were fifty years old. In that period, the State had awakened to the need for development in the oyster industry and had established by plantings many new and highly productive bars, most of them extensions of old grounds, and none of them shown on the official charts.

One other factor must also be considered. Oyster tongers are permitted to work only in the county in which their licenses are issued. Clammers, uncontrolled by law, crossed county lines at will. These experienced and intelligent watermen soon learned that the Miles River produces abundant clams of excellent quality, and, in a very short time, the river floated every rig from Queen Anne's County and some from as far away as Anne Arundel. We had them all, including Talbot rigs from Tilghman's Island, Oxford, Bosman, and Neavitt.

So the tongers watched in helpless fury while the clam dredgers annihilated one oyster bar after another. They destroyed Gibson's Flats, pumped up Newcombe Hollow, and devastated the fine bars of Leeds Creek. And the same wholesale destruction was also taking place in the Wye River.

But what could the tongers do, or the police either? These bars had no official existence. When you think of it, it is amazing to what extent we live by, for, and on paper. As for the tongers, there was talk of burning boats. When thoroughly aroused, tongers are quite capable of burning anything. But some of the boats threatened belonged to their own brothers, fathers, uncles, and cousins, or to men with whom they had worked and whose company they had enjoyed since childhood. The tongers were helpless.

That the clammers knew what they were doing cannot be denied. Captain Will Roe came alongside a dredger working on Gibson's Flats. He wanted to buy some broken clams for eel pot bait. The clammer had eight bushel baskets of oysters stacked neatly away, one on top of the other, in his cabin. But the get-rich-quick fever of a new industry burns away old restraints and sends old standards up in flames. The big grab was on.

To make a bad situation even worse, the inshore limits of many of the official bars (those entered on the old charts) were in many cases ill-defined. The temptation to edge out further and further from shore and onto the bar proper met with little resistance. I watched twenty-four hydraulic clam dredges work week after week on the shoal areas of Long Point Bar and permanently reduce by at least fifty per cent the productivity of one of the finest oyster grounds in the upper Miles. And the same thing happened on Maiden Point.

Since the legal situation was confused to the point of chaos, it was almost impossible for the police to get convictions in the courts. Emboldened by the apparent helplessness of the enforcement officers, the clammers now flagrantly dropped their scoops everywhere and anywhere. Only the most energetic efforts on the part of the Marine Patrol saved the remaining bars in the Miles River. Already the clammers were introducing longer scoops for work in deeper waters. Little by little, however, the police brought the worst offenders under control. Convictions were now gained against the most flagrant of the violators. I think it is interesting that, in 1953, sixty-five clam rigs paid a total of $6,350 in fines, and this in an industry virtually uncontrolled by law.

The tongers howled for help, but "wiser heads," that

9 EELS TRAPPED IN AN EEL POT

10 EELS IN EEL POT

11 EELS BEING DUMPED INTO BARREL FOR SALTING

12 HYDRAULIC CLAM DREDGE IN OPERATION

is, men who had never worked on the water in their lives, assured them that the hydraulic dredge could not possibly do as much damage as was claimed: in fact, it probably did only negligible and reversible damage, if any.

Finally, in 1956, the Department of Research and Education acquired a clam dredge and with it determined the amount of damage done when this rig was used on ground bearing oysters. In the dredged area itself, it was found that heavy oyster mortality occurred. Within seventy-five feet of the dredge, on either the ebb or the flow side, depending on the tide, heavy to light mortality was observed.

By that time, however, the harm was done. But even before these tests had been made, news of devastation in Talbot and Queen Anne's counties had spread rapidly over the entire tidewater. In 1955, the General Assembly passed an act establishing general control over the use of the hydraulic dredge in the harvesting of clams. The minimum legal size was enacted as two inches, and an annual license fee of fifty dollars was charged for every rig. As a reflection of the bitter enmity between the tongers and the clam dredgers, the same act prohibited the use of the hydraulic dredge within five miles of any oyster bar in the waters of Anne Arundel County! This act also prohibited the use of the dredge in any of the waters of Somerset, Kent, St. Mary's, Dorchester, and Wicomico counties.

Later legislation has also limited the size of the scoop-head and its length, as well as the amount of water pressure allowed in the jets. If the law had not also prohibited the use of the dredge within fifty feet of any established structure in the water, there is no doubt in my mind that the clammers would have dug up every duck blind, bridge, and wharf in the county.

But the clammers formed an Association and established a lobby. Probably its most powerful argument is its ability to make money, the usual weapon of a lucrative, young, and growing industry. At any rate, the lobby has been successful.

In 1958, the very county that had, in 1955, prohibited the use of the clam dredge within five miles of any oyster bed, now made an important concession. A portion of Three Sisters Bar in Anne Arundel was resurveyed and examined for a concentration of oyster cultch and soft clam populations. As a result of these surveys, 330 acres of this bar were reclassified and opened to clam dredging. And in Queen Anne's County, clamming areas have been extended along Kent Island's shore.

In 1958, clamming was permitted by law in Anne Arundel, Calvert, Queen Anne's and Talbot counties.

Conservation

The damage has been done, and there is no point in nursing old grudges. The division between tongers and clammers has been further healed by controlling legislation adequately backed by enforcement and by the local courts. The two industries, the very old and the very new, now operate side by side with little friction.

The vision of fortunes to be made from the exhaustless riches of unfailing clam beds has vanished like the impossible dream that it was, and the clammers now take a more sensible and a more responsible view. However, they still have much to learn as far as the conservation of clams is concerned. For one thing, there is still no closed season, and maninoses are dug summer and winter.

During its youthful and irresponsible period, the Clammer's Association circulated a piece of propaganda which it may have persuaded itself to believe. The claim was that the hydraulic dredge actually cultivates the bottom and so aids in producing unlimited supplies of clams for succeeding years. This piece of pie in the sky has long since proved to be inedible. The hydraulic dredge, like the oyster dredge, is a very efficient tool, and it can dig clams much faster than the rate of replacement can possibly restore them.

When a maninoser works his rig, he picks from the moving belt the market-size clams. The others run off the end of the belt, drop overboard, and sink to the bottom. It is now known positively that the very small clams quickly dig themselves back into the bottom again. The larger clams, however, are unable to re-establish themselves in the sand and soon perish. Now this is pure waste, and the clamming industry, if it wishes to survive, will have to solve the problem.

The need for conservation is all the more pressing when one considers that all of the available clamming grounds have now been worked over many times. The virgin crop has been harvested and the big grab is over. Now the catch per day per boat has declined, and it will continue to do so. As a result of this reduction in catch, clammers are pressing for new grounds. But if they get new grounds, the same process of attrition will again occur, and very rapidly. In the end, conservation, and only conservation, can save the clamming industry from self-destruction.

For a species called *sapiens,* it takes us a long, long time to learn.

fishing

5

Angling

I admire the fish hawk for his skill, his strength, and the beauty of his flight. But my admiration is not unmixed with envy. He catches fish. I seldom do. To this day, any twelve-year-old in Tunis Mills can put me to shame with hook and line. As for trolling, the only time I can catch rock is when they school and break the surface; even then, the stripers have to be hungry fish.

It is an exciting thing to see a big school of rock break. They beat the water to foam, while minnows and alewives leap in all directions as if flung from the water by the handful. Blues are even more savage in their hunger. After the passage of a school of blues, I have often seen dozens of alewives floating in the water, but only one half, either head or tail. The rest, presumably, remained in the jaws of a blue. Blues have small teeth like those of a kitten, just as sharp as needles.

But whether blue, or rock, or hardhead, I have had little luck, though I will say I can catch dowdies with any man. I should add that a dowdy is a toadfish, said to be entirely inedible.

Gill Net

Haul seines are used in the Chesapeake; so are pound nets and fykes; but the most characteristic net, for both commercial and amateur fishing, is the gill net. Commercial fishermen use this netting in different ways, and it goes, therefore, by different names. The anchor net is simply a gill net which uses buoys and anchors instead of stakes to hold it in place; a drift net fishes deeper water and is allowed to drift across the bottom; and a stake net is one held by stakes driven into the bottom. But they are all gill nets. In 1957, licenses were issued by the State of Maryland for 599,028 yards of gill netting. In Talbot County alone, 57,360 yards were used.

Not only is the gill net commercial gear, it is also used by amateurs to supply the table. A non-commercial gill-netter is allowed to fish one hundred feet of net without a license.

The gill net works this way. The mesh is made of linen or nylon twine, and it is about five to six feet in depth. There are corks every six or seven feet to float the net, and iron rings at the bottom of the net opposite the corks at the top to make the net stay straight up and down. Stakes at either end keep it stretched out in place. Fish come along at night, and they are unable to see the fine linen mesh. They shove into it head first, then shove again violently when they feel resistance. The fine thread or twine pushes back over their heads. When the fish decide to back out, the twine of the mesh gets caught in their gills, and they can go neither back nor forward. Your fish is gilled.

When I was a young fellow, I decided early one summer to catch fish, not just go through the motions. My brother procured me some fine linen mesh, corks for floating the net, and rings to keep the net hanging vertically from the corks. Having hung my gill net to cotton ropes, I decided to catch all kinds of fish and plenty of them. I determined to eat little or nothing but fish for a month or two, because I was as hungry for fish as a schooling rock.

I distinguished myself with that gill net. I do not suppose any inhabitant of the village of Tunis Mills, living or dead, has ever had such hard luck with a net, except one, and I will tell you about him later. Of course, summer is not the best season for gill-netting. I realize that. But even so, my brother caught a few nice rock, a few hardhead, and some perch with that same net and in the same time of the year.

A gill net is set out at dusk and left to fish overnight as a rule. But in the summer time, crabs will get entangled in your net. They keep gathering and gathering the meshes with their legs and claws, at the same time chewing the twine, and what they can do to a gill net is sad to relate. So you have to stay by your net and shake the crabs out once in a while. If you leave them in the net, it is impossible to get them out unless you break the twine. The only possible method of removal is at once primitive and effective. You put the crab with the bundle of linen meshes in the bottom of the boat or on the stern seat or washboard. You then step on the crab repeatedly until he is reduced to pulp. The pulp can then be shaken from the meshes. It is a waste of time and crab meat. In addition, the more sensitive might possibly find the process a shade on the repulsive side.

Well, I stayed by my net, night after night. Religiously, I overhauled it every hour or so. I caught a couple of white perch one night. Every night I caught crabs. For weeks previously, I had searched for crabs around the shore without finding any. I could have sworn there was not a crab in the creek. Every night I caught crabs, many crabs. The net was suffering. I therefore decided to try a new location. I had noticed an occasional big white perch jumping in the water above the bridge, and I was fairly sure there were few crabs at that time of the year so far up the creek. I left the net out all night.

By this time, I was getting more than a little sensitive to the attitude of my neighbors, most of whom had followed the water at one time or another, all of them interested in my netting operations in a friendly, but at the same time, critical way. They let me know in a courteous fashion that I was an amateur and that, aside from the fact that I could row and talk, I had everything about gill-netting still to learn.

Jess and his brother, Les, would see me starting out from their shore about sundown, and one or the other would say, casually but with an eye-glint of humor in otherwise impassive faces, "Well, cap'n, I guess you loaded the boat last night?"

What can a man answer who has caught one dowdy, one alewife, and three crabs? On one occasion, I was so flustered that I left the tub containing the net behind me on the shore and had to row all the way back for it. Jess waited for me with infinite amusement.

I was burning to make one good catch, just one. It was now a matter of pride, and Satan himself can sin no further in that direction than a crabber, tonger, or fisherman. The pride takes the form of offhand modesty in regard to a very good catch. It is a sweet moment in life to put out ten bushels of oysters, for example, and be extremely casual about it: a pretty good little jag. What I lived for was one of those pretty fair hauls, say a whole damn net full. I had to have it.

I slept the night I left the gill net above the bridge, but not without dreams. I caught everything in that net. I caught rock and blues; I caught snapping turtles and diamond-backs; I even managed to catch two fish hawks and a coon. I caught everything except what was really in the net when I fished it early in the morning.

At first, when I rowed to the inshore stake, I thought there was nothing in the net at all. Then a cork out toward the middle bobbed under. I watched it. It bobbed under again, way under. I held my breath, It reappeared, but not until after an appreciable length of time. I said, "Son, you've caught a big one. God, he must weigh eight pounds or more."

The meshes inshore held nothing but a small hogchoke and a horseshoe crab, both of which I shook out with no trouble. Anyhow, I had no eyes for anything but that deeply disappearing cork in the middle of the net. By the time I reached it, my hands were trembling with excitement.

With a big fish, you have to be a little careful, for he may not be securely gilled. Pulling up the net may give him just the tension that he needs to escape. So I used a trick my brother had taught me. I gathered in the meshes on either side of the critical cork and got hold of the bottom of the net. That way, I had the fish in a sort of linen cradle from which he could not escape. I lifted the cradle gently, peering down the while to see what manner of fish the waters had given me. I caught sight of a big, struggling, chunky body, and thrilled with excitement. I could feel the weight of him now on the net, and the tugging as he threshed.

I have never, before or since, known such an instant of utter amazement and blank incredulity as I experienced when the fish and I came literally face to face, for I was bent far over the skiff with my eyes as close to the net as I could get them. My fish was a muskrat.

He was neatly caught by a loop of linen mesh about his off hind foot. He looked directly into my eyes, drew back his split lip, and clicked his pointed teeth together in fear and anger. My fish was an eight pounder, all right, possibly more.

If I had known then what I found out later, I might not have let him go. As it was, I drew a knife and, being careful to avoid his teeth, snicked the single twine that held him. Here he had come swimming along, minding his own food business, and had suddenly run into a thing which yielded but would not release. He must have spent most of the night swimming to the surface, carrying the weight of the net with him, to get a gulp of air; then the struggle again to free himself under water from the soft, flimsy, yet bitterly binding thing that held him. I let him go.

Having disposed of the rat, I proceeded to fish the rest of the net, which, at one time during the previous night, had contained seven large white perch, judging by what was left. Heads were what was left, neatly gilled heads, and nothing else. It was now plain to me how the muskrat had gotten caught. He and his brothers had fished my net for me, and very efficiently. I suppose I could have taken seven fish heads back with me to show Les and Jess that I really had caught that many fish. But who wants to be the laughing stock of a village? It was bad enough as it was.

I was not licked yet, however. I would try the Miles River, where the big stuff was; get out of the creek entirely. That very night I put the net out just before the turn of the tide, a nice set. As soon as the tide turned, the corks began to bob. I caught dozens of fish, all alewives, the biggest and fattest I have ever seen, and all useless, for Shoremen have an invincible prejudice against these fish. I understand that they were caught during World War II and canned in large numbers, and that the fish so prepared were as good as herring. But I would never eat one fresh myself, nor would any Talbot man I have ever met.

So far as I know, these fish have never taken a hook. Their apparently aimless wanderings in small schools close to the surface of the water have puzzled me since I was a boy. I could not see how they fed themselves, how they continued to exist in such prodigious numbers, always a prey to larger fish. I now know. They feed exclusively upon plankton, their narrow mouths being especially adapted to food intake of that kind. In order to feed, all that they have to do is swim. They spend their entire lives in ceaseless motion. I have read that this species of herring is called alewife because it has a larger belly than that of the true herring. All of you who are true malt worms now understand how the alewife got its name. In Talbot County, however, the name has been corrupted to L-Y. For many of my younger years I enquired restlessly of the villagers why an L-Y was so called. The only answer I ever received was that the gill case formed an L and the tail, a Y. Since that is roughly true of most fish, I failed to find any satisfaction in the answer.

When I had taken alewives out of the net until I was slobbered with slime, I took it up in disgust, resolved not to put it out again until I could try the spring run. Like better men before me, I broke my resolution. A small, but very active school of rock invaded the creek not a hundred yards from the shore where I kept my boats. What is more, they were seen by Les and Jess, both of whom urged me to get them if I could. I saw no reason why I could not. It is true that they were feeding close to shore in heavy grass, so that trolling was out of the question. But, after all, I had a gill net.

The matter struck me as being comparatively simple. Watch their location at dusk. If the tide was going out, set the net below them; if coming in, above. The result must be a whole netful of small but eminently edible striped bass, ranging from twelve to sixteen inches in length. I tried. I tried every night for a week. Those exasperating fish tore the water to pieces not fifty feet from the net on two occasions. I never caught a single fish. I do not know why I continued to set, but I did. I suppose it had gotten to be a habit by that time. The whole village, of course, had seen me go after that school of rock. By the end of the week, I found myself avoiding people. When someone called my name, it was all I could do to keep from wincing. If boats came by in the dusk past my set, I turned my head in the other direction and pretended interest in a persimmon tree on the creek bank. Some other tree would have been more appropriate: quince, say, or crab apple.

On the last evening, when the sun had set, and a clear flush lingered in the deep sky, a boat hesitated with throttled motor at the outshore stake and finally stopped. I looked sedulously at the persimmon tree and waited for the inevitable, good-natured, but acidly ironical question, "Well, cap'n, goin' to fill the boat tonight?" Instead of that a mild and soothing statement of fact was murmured.

"Got a nice set there, a nice set, certain."

I turned to the speaker. He sat on the washboard of his workboat and held the net stake with one hand. It was Captain Charles Draper, a man who had fished and oystered before I was born or rumored.

"Never be afraid of a bow in the net when the water's warm. A slack net'll catch more fish than a tight one. 'Course, when the water's cold, they gill clean and don't fight the mesh so bad. I see you have her across the tide. That's right."

"I wouldn't know, Cap'n Draper; I haven't caught any fish to speak of in warm water. But I'm glad you like the set."

"No luck?"

"No, sir. No luck."

"There's luck and there's luck. What kind do you want? A netful, a whole school of rock?"

"Yes, sir," I said defiantly. "I want a school, a whole damn school.

"Ain't that somethin'? That's what I always wanted too."

"You got a whole school once?"

"I know it's so."

I kept very still and quiet, in spite of the gnats that burned my hands and face like fire. They had no effect on Captain Draper; they were afraid of his old pipe, seldom cleaned and always used. I wanted to hear about a big haul, even if I couldn't get one myself.

"This was before the yellow perch went," he said. "Remember?"

I nodded.

"Well, I was busy. I was tongin', I didn't have no time, to speak of. But I knitted me a gill net, three, four times the length of that one of yours. Every time it snowed or blowed, I worked on that net. Says to myself, I'll have that net ready by April; set it out in a cove over Wye way I know about; full of yellow perch. I'll eat some and I'll sell some. Got no time to fool with rock, not till after oyster season. Just set her there and leave her. Fish her onc't a day.

"When April come, I had that net knitted and hung. I never made a better, only some might have thought the linen a little light, for I favor a fine thread. I caught some perch, too. Then, later on, in June it was of one Sunday evenin', I thought I would try for some rock. I picked my ground and set her out, a little tighter than yours, 'cause I wanted to cover as much bottom as I could. Maybe too tight. Well, it was a fine evenin', one of them mild, bright ones, just like this. So I figured, why not stay here a while at the stake, I got nothin' else to do? So I stayed. And the longer I stayed, the more I got to wonderin' how it would be if a whole school of big rock come adancin' and ajumpin' up the river and into the cove and right smack into my net. A whole damn school. Yes, sir.

"It was one of them calm evenin's like this, sundown,

13 HYDRAULIC CLAM RIGS AT WORK. NOTE SILTED WATER FROM HYDRAULIC PRESSURE AND HOW SILT STREAMS AWAY ON THE TIDE

14 THE AUTHOR GILL NETTING

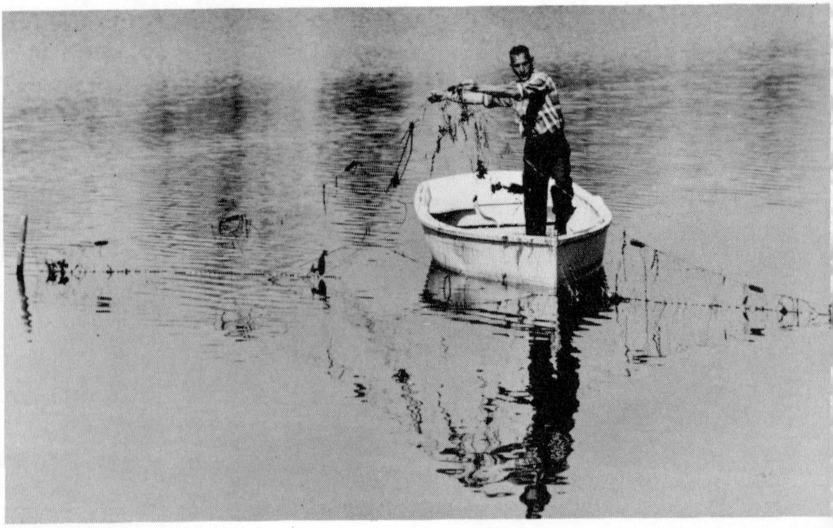

15 GILL NETTING

but a lot of light in the sky, the kind you can see five miles in and hear ten. And about that time, away down by the mouth of the river off Bennett's Point, I seen somethin'; and I heard somethin'. You've seen it and you've heard it. A damn big bunch of rock just tearin' the water up. The tide was flood and them fish was movin' with it, workin' right up the river, real slow, till they come right opposite the cove. And I prayed. I said, 'Dear God, make them fish come up the cove.' And they come up the cove.

"I sat there holdin' on to the outshore stake just like I'm holdin' this one. And I was as stiff as a wet shirt in a freeze. All I could do was sit there and watch them fish atearin' the water and adrivin' the L-Y's clean out and into the air as high as my head. They come on and they come on. And they hit my net, a thousand of them, all big, about five, eight pounders."

For a while, there was an awed silence on my part, a contemplative one on Captain Draper's. I could stand it no longer.

I said, "How many?"

He chuckled. "Not one. Not a single one. They tore that net to shreds and pieces. I didn't have nothin' left but a few corks and two stakes. You ever catch a rock with a linen mesh around his head, he's mine, and don't you forget it."

There was another silence on my part, equally awed, while Captain Draper knocked his pipe out and went aft to start his motor. Before he did, however, he looked up and said, "We all get gilled, fish and man alike, soon or late."

After he left, I took up my net and it stayed out of water for the rest of the summer. Aside from Captain Draper's story, the sea nettles were getting bad. They hang in the net and burn your hands every time you overhaul. The acid they emit on being touched will rot your twine if the net is not thoroughly flushed with fresh water when it is taken up. What with crabs, nettles, and Captain Draper, I had had enough for a while.

I won't say I disbelieved the Captain's horrible experience, but I wanted corroboration. All the Tunis Mills men who were about when Captain Draper lost his net agreed that he went out with a good, new gill net one June Sunday evening and, two hours later, returned without one. Nor was the net ever seen again. In the bottom of his boat when he returned were a few corks and two net stakes, nothing more. When asked if they put full credence in his story, they replied that he certainly did not *throw* a good net away, for, as everybody knew, he was as tight as a nail driven to the head in oak. I leave it to you.

There is one thing in the good captain's story, however, which I happen to know is a fact. The yellow perch did disappear; or, as he put it, they went. Everyone says it that way, although these fish did not exactly go. They were joyfully assisted on their way. Their disappearance seems to me to be an interesting example of the local depletion of a species. Captain Draper, by the way, took a major part in speeding the yellow perch.

One very cold March morning, a late freeze set in after mild weather. The water, which had been somewhat warmed, refused to ice solidly. It froze in a kind of loose but heavy slush. It may be that a half gale of wind agitating the water aided in this curious formation of loose crystals. At any rate, the yellow perch, caught in their annual spawning run, suffered severely. Had the water frozen normally, they would have remained comatose but safe under several inches of ice. The first thaw would have seen a return to their usual activities.

Mr. Arthur, alive then and dwelling in his house across the lane from me, made his usual early morning visit to the bank of the creek and saw a strange sight. Hundreds of yellow perch floated in the heavy slush formed by the late freeze. He saw at once that it would be possible to launch a rowboat, and, to a degree at any rate, maneuver it, a thing it would have been out of the question to do, of course, had the water frozen as usual. Aside from will, hardihood, and oars, the only other equipment needed was a dipnet. All were available. Mr. Arthur got his boys to the shore, launched the rowboat, and proceeded to do some of the strangest fishing these waters have ever seen. In a comparatively short time, the boat was filled and the perch "went."

Captain Draper, working his oyster boat out of a small creek emptying into the Miles River, noticed that his keel was plowing through fish and slush. He had a dipnet on board. In a matter of an hour or two, he was standing to the top of his sea boots in yellow perch.

So the yellow perch have gone. On the other hand, so has Captain Draper. As he pointed out, we all get gilled, man and fish alike, soon or late.

Fish Kills

Fish kills occur either for natural reasons, as in the case of the yellow perch that went, or for causes directly attributable to man's activities on or near the Bay.

In the last week in May of 1955, large numbers of fish of many kinds were seen either dead or dying from Tolchester Beach down to Kent Island, a distance of fourteen miles. Most of the fish were white perch, but a great many other species were observed, including eels, yellow perch, black bass, and menhaden. Some beaches were so littered with dead fish it was difficult to walk without stepping on them. So far as could be determined, this kill resulted from suffocation when large amounts of vegetation exhausted the oxygen in the water.

During the summer, most of the Eastern Shore shoal waters are choked with sea grass. It is so thick in some areas that it is quite impossible to get a boat through it, due to the strands fouling the propeller. And a crab line laid in such grass can not be worked and is very difficult to retrieve, since tons of grass catch in it and produce an anchoring effect. This broad-leafed variety dies in the fall and rolls around on the bottom all winter, driven by tides and storm seas. But most of it collects in great masses near shoal water. When the water warms in spring, these tons of old grass begin to decay rapidly and draw off the local supply of oxygen in the water as part of the process. I have known these mats of dead grass to be so thick on shoal oyster grounds as to make tonging impossible.

Although some fish kills are known to be the result of natural causes and are, to a great extent, quite unavoidable, others are caused by industrial and domestic pollution, or are the aftermath of insect control by various chemical sprays. Since many cultivated crops are sprayed several times a season, these chemicals undoubtedly kill many fish each year. What happens is that heavy rains carry the poisons into the water.

Bilge pumping by freighters passing through the capes and up the Bay to the port of Baltimore is another cause of fish kills. The practice is forbidden by law, but freighters continue to violate it. During World War II, the harbor of Philadelphia was so impregnated with filthy, oily bilge pumpings that the air (by courtesy) on and around it was very nearly unbreathable. During that period, I had occasion to cross by ferry from Philadelphia to Camden, and it was impossible to stay on the decks. The passengers, myself included, herded together in the sweating heat of the lounge, every window tightly closed against the miasmic stench of what, by a wide stretch of the imagination, might have been called water.

A large oil slick was found by the Tidewater Police in the upper Bay in the spring of 1956. It was spread over the same area and at the same time rock fish (striped bass) were spawning. The violating vessel was not found, but the Department of Tidewater Fisheries, to prevent such dangers, has increased patrolling of spawning areas.

The Catch

There are over two hundred varieties of fish caught in the Bay in summer time. Of these, all but striped bass, white perch, and eels leave the Bay every winter. Since the decline in abundance of bluefish in the Bay since 1937, the fish of highest sporting and commercial value is the striped bass. Commercial fishing is limited by the number of licenses issued for nettage each year, and the catch has remained relatively constant for many years. For example, the commercial catch for the year 1944 was 13,873,017 pounds; for 1953, 13,176,935; and so it goes, year after year.

The Feud

For the last ten years or more, a conflict of interest has developed in the Chesapeake Bay until it has now reached the proportions of a perennial feud between sport fishermen and those who catch fish for a living. Sportsmen are strong lobbyists and extremely verbal partisans of legislative action favoring themselves and limiting the activities of commercial fishermen. Since sport fishermen have received wide support and publicity and have had their case stated again and again, year after year, I think it is about time that someone said something in defense of commercial fishermen.

First let me state one obvious but basic fact which is often overlooked, and then we can get onto other aspects of the matter. If you have ever fished a big gill net in a March sleet storm, you will be convinced that you are working for a living. And that is precisely what commercial fishermen do. They work for themselves and for their families, and, if necessary, they half-kill themselves to do the job. The sport fisherman is merely killing time, not himself; he is just fooling around. Let him fool as much as he wishes, but it is not for him to grudge watermen their living, which is won hardly in all weathers and all seasons by men who have a vital stake in the resources of the Bay, not merely a casual and a passing interest.

While commercial fishing has remained constant for two decades or more, sport fishing has grown dramatically until it has reached the proportions of a big business. Millions of dollars are spent by sport fishermen each season on food, boat hire, beer, whisky, equipment, and lodging. They had already attained such large numbers in the Chesapeake Bay by 1953 that their total catch of fish was estimated by the Maryland Board of Natural Resources at 3,388,000 pounds, which is probably far under the actual take. Now, the commercial catch for that year was 13,176,935 pounds. If the sportsmen's take is added, we have a total of 16,564,935 pounds. So sportsmen caught approximately one-fourth as many fish as did commercial fishermen in the Chesapeake and its tributaries, and one-fifth of the total catch, which is not bad for hook and line. Sport fishermen had nothing to complain of, it seems to me, nor have they now. They are still getting a very large share of the total catch.

The various outdoors columns in the newspapers and sporting programmes on radio and television have been devoting increasing space and time to salt water fishing, and through their efforts the most productive areas and the best and newest methods for taking fish of all kinds have been widely publicized. The chances of catching fish are now better than ever, as far as the sportsmen are concerned. And yet they continue to put pressure on the General Assembly for changes in the tidewater laws favorable to the angler.

In 1955, haul seiners on the Susquehanna Flats had good luck with rock, and sizable catches were made there shortly after the middle of July. Rock over fifteen pounds, which are protected by law, at times outnumbered fish of legal size in the nets and presented a difficult culling problem for both the fishermen and the law enforcement officers who supervised the catch. The large catch landed and the loss of oversize fish so enraged envious sport fishermen that their lobby pressed for a measure to close the Susquehanna Flats to haul seiners, an area which had been successfully fished by them for many years. The lobby was successful, and, in 1957 legislative action closed part of the Flats to haul seiners.

In the same year, the minimum legal length for rock was raised from eleven to twelve inches. While commercial fishermen had to comply with the new legal length because of strict supervision, sport fishermen were getting away with murder. It is impossible to supervise thousands of craft and unlicensed fishermen whose catch is unmarketed, at least in the usual channels. I have more than once seen twenty-four-inch galvanized garbage cans, filled with undersize rock, taken ashore and carried off in the back of sport fishermen's cars. It was reported by tidewater police that in some areas half the fish taken by anglers were under the twelve-inch legal minimum, and I have myself seen numbers of rock over the legal maximum caught and kept by sportsmen.

The Tidewater Fisheries Report for 1958 makes, in reference to sport fishermen, what must surely be the understatement for that year:

> Considerable time was spent by the enforcement staff in obtaining adherence to the law.

And yet the angler continues to complain vehemently that he is being wiped out of existence by those nasty men in boots who fish for a living. It is not as if the sportsman never catches fish. He does.

A sport fisherman trolling in Tangier Sound the first week of November 1955 took a hundred and sixty-eight rock in a single day, some of which weighed up to fifteen pounds. In other words, he got about half a ton of fish.

Toward the end of July of the same year, a party of four caught two hundred and eighteen rock on Black Walnut Bar at the mouth of the Choptank River.

In 1957, a party fishing Eastern Bay in November landed between four and five hundred white perch in a single day. In the same year, I saw a four-by-four-by-three-foot box completely filled with rock, the product of an afternoon's sport and the rods of two fishermen. These were fine fish, weighing two to four pounds apiece.

Catches such as these are by no means isolated instances. Big hauls are frequently made by anglers and trollers all over the Bay every summer; and the sport fishermen have extended their season. Each year, they start earlier and stay later, sometimes too late. Three of them, along with the captain of their party boat, were drowned in a northeast snowstorm in December of 1960. In December of 1958, many big catches were made between Bloody Point Light and Point Lookout.

As the sport fishermen increase in numbers and gradually extend their activities into early spring and late fall, the catch of rock by commercial fishermen declines year by year. After all, the angler is also looking for striped bass.

In 1954, the commercial catch was 14,946,422 pounds. It was valued at 1.5 million dollars to the fishermen. Due to increased yields of menhaden, carp, and spot, the total poundage for the year was eleven per cent greater than in 1953. But landings of striped bass, the industry's most valuable fish, showed an eight per cent decline.

In 1956, approximately the same poundage was landed, but there was a further loss in fishermen's income to 1.3 million dollars because of increased catches of low-priced fish. The striped bass catch was seventeen per cent under that of 1955, and ten per cent under the average annual total in the ten-year period from 1944 to 1954.

Again in 1957 commercial fisheries suffered a further loss of income as the result of a drop by 300,000 pounds in the landings of striped bass. There is no doubt in my mind that the commercial catch of rock and the fishermen's income will continue to fall as sport fishing and restrictive legislation increase.

Other factors may also be involved in this continuing loss, but it is obviously not the commercial fisherman who hurts the angler. The shoe is on the other foot, and it seems most unsporting of sport fishermen to make every effort to limit and harass men whose work is highly useful and whose living comes the hard way.

Indeed, it would be much more to the point if pleasure boatmen in general examined their own conduct on the water and made a firm effort to improve it, for it certainly needs a thorough overhaul.

More and more people are taking up residence on or near the Chesapeake Bay. In addition, the mass production of small boats, the increased use of trailers to carry them, and the improvements in outboard motors, have brought what can only be described as a period of swarming which begins in April and extends well into November. Watermen in every kind of job have had their tempers sorely tested by thoughtless and ill-mannered pleasure boatmen. I, for one, dread the thought of tonging on or around Labor Day when the great swarm reaches a peak of frenzied activity. I do not at all mind being watched if I am allowed to get on with the job. Let them stand off and glue the glasses on me; it is all right with me. But when they see the quaint little man standing on the washboards with a pair of long sticks in his hands and then blow by not fifty feet away at thirty miles an hour, the quaint little man wishes to hell he had brought a rifle along. For days on end, it is almost impossible to stand on the washboards, and you find yourself longing for the first hard frost and the bitter northwest gales, which seem delightful by comparison.

They also clear the water of pleasure boats.

Trot-liners have the same tale to tell, and pot crabbers are finding it difficult to make a living. Their buoys are often deliberately cut by pleasure boatmen, who find them a hazard to navigation. The buoy once lost, the pot and its catch are also gone for good.

Again, it should be remembered that pot crabbers are trying to make a living for themselves and their families in a hard and useful job that helps to feed thousands of people. Presumably, boatmen are on the water because they enjoy being there. In that case, they should be quite willing to take the little additional time necessary to run through a group of pot buoys. That is all it takes, just a little more time and care, a decent desire not to cause loss and hardship to men who have not put their pots in that particular place to cause a minor annoyance but solely and simply because crabs are to be found there; and to make a living, they must find crabs. In other words, all that is needed is the most common and garden variety of courtesy, an article which we have not been noticeably successful in mass-producing.

Some damage is even more wanton and malicious. Crab pots are often fished and then completely destroyed. Many pot crabbers have petitioned the Tidewater Fisheries Commission to allow the setting of pots on a line with only two markers, one at either end. This is not the most efficient way to work these pots, but the hope is that they will then be less frequently disturbed. Actually, the only hope lies in an improvement of behavior among pleasure boatmen.

I do not in the least grudge the sport fisherman his fun. But, speaking for myself alone, his increased numbers and lack of self-control have made trolling a chore rather than a pleasure. As all fishermen know, one of the signs of a fish school is a big flock of sea gulls hovering over the water in the hope of picking up minnows and alewives driven to the surface by rock or bluefish. As soon as gulls are sighted over feeding fish, boats by the dozens roar in from every point of the compass and drive the fish right under. Having made the fish sound and spoiled the sport for another hour or more, the boats then begin to go round and round in a large circle, stem to stern. The smell of diesel fumes and gasoline exhausts sweetens the summer air, as half-eaten hot dogs, lunch boxes, wax sandwich wrappers, pickle jars, and beer cans float in the water, lending an atmosphere of happy picnic informality.

On May 25, 1958, more than seventy-eight boats were counted fishing on Brickhouse Bar off Kent Island.

Thanks very much, but no thank you. For me, fishing is a quiet and a lonely thing.

hunting

6

Crab Net

I am not a hunter myself, and I think I can prove it. I used to own a shotgun, but I sold it to Mole Daffin, who is a hunter. That gun was going to rust and dust while I owned it.

I had trouble one spring with a rabbit, a big fellow. It seemed to me that all my hard work in the garden was for one wild rabbit who was getting just a little tamer every day. After a while, he made himself right at home, never left the lot.

The lettuce would just begin to head up, a fresh, pale green with a fine little curl to every leaf, and that rabbit would go right down the row at night. In the morning, the early sun would sparkle on a new, green world, all but that row of lettuce. Nothing left but the stumps, with the juice already turning a dirty brown. Murder by night.

And the peas began to come on well, perfect miniature vines with a tiny tendril already uncoiling at the top of each plant. The rabbit didn't even leave the stumps this time. Right down to the ground. You couldn't even see where all that promise had been. I'm very fond of peas, too.

A hunter wouldn't have had my trouble. He would have shot the rabbit, if he had had to borrow a gun for the job. All I did was swear, and stomp, and throw oyster shells at the beast, who would hop off in a fat, lazy way. He would stop after a token run, dip his front paws in the dew, and wash his face in the most unconcerned manner possible.

"Oh hell," he says, "what's all the fuss about? A few peas, for Christ's sake."

He knew a contemptible, non-hunting animal when he saw one. From that moment, his character degenerated visibly. He developed a suicidal overconfidence and contemptuous pride. *Hubris* was his fatal flaw. He began to eat my wife's anemones.

Every gardener will know what I suffered from that rabbit. And every married gardener will understand that the combination of an angry wife and a ravenous rabbit is simply overwhelming. Under these pressures, the most humanitarian of men will feel a need to kill. With a less balanced individual, it is sometimes the wife who is an object of this sudden need for blood. But I hope I am a reasonable man. On the whole, it seemed to me it was the rabbit that must go.

Did I do the manly thing, go and borrow a gun and shoot the rabbit? I did not. I did what only a non-hunting waterman would have done. I captured the rabbit in a crab net. This flaw of character I have already mentioned, this o'erweening pride, helped materially. He just sat there and I dropped the net over his head.

I requested my wife to stand on the net handle, and I went off to find a neighbor, Jack Soulsby. He had not long finished his breakfast.

"A rabbit? Where?"

"In my yard, there. In a crab net."

"A rabbit in a crab net," says Jack. "I've heard it all now. Early in the morning for jokes, you ask me."

I told him all over again. He smiled indulgently.

"All right. Let's go see the little rabbit in the crab net."

"Yeah," I said, "and just in case the little rabbit happens to be in the little crab net, suppose you tote your little .38 along."

So Jack killed the rabbit for me. Did I sense in him a portion of the same good-natured contempt his victim had shown for the non-predator?

You see how it is. When it comes to blood sports, it will be necessary to let others do the talking.

The Law

For five summers, I left the water and worked ashore. Soon after hanging up the oyster rakes in April, I went to work with a local construction gang. I did that for two summers. It was hard, hot work. So the next summer, I got a job in St. Michaels harbor at a boatyard.

The nature of my work was such that there were slow times as well as busy ones. During a hot, windless afternoon in August, I sat on the bench in front of the boatyard store and listened to Mr. Med and Greenie swapping yarns.

Mr. Med was the owner of the yard, which he later sold. He was a famous sports fisherman, well known on the Miles River, but now he had scarcely the time to wet a line any more, he was that successful in business, which is still another instance proving that success simply does not pay.

Unfortunately, Greenie also had a bustling business in the town of Easton, so the pair of them looked forward to slack winter times and to duck hunting.

You will have to picture Mr. Med as a man of about sixty years of age with a greasy leather cap pulled down to protect his eyes against the blinding glare from the harbor waters. He stands four square, short but solid and still muscular, an able mechanic, an able man. He has a somewhat used look, like that of a person who has lived, played, and worked as hard as he could. Tough and self-confident, hot-tempered and humorous. Greenie is younger, with blond hair and light eyes, handsome and completely relaxed on the store bench. He is in the summer of his prime, powerful and indolent at the moment, but a man of great energy, nevertheless, ready for anything. But more of Greenie later.

Mr. Med hitches up his grease-stained trousers, cocks his leather cap.

"Greenie," he says, "either a law is a good law, or it is a bad law. What makes a good law?"

Greenie laughs good-naturedly and says, "Hell, Med, I ain't but a grease-monkey like you. How would I know?"

Med takes his cap off, looks at it disdainfully, and slaps it back on again.

"Was I to be your age, Greenie, and not know more than you, I'd hide away under a bulkhead somewheres and have them bring me my meals. Listen. A law comes out says you can't use a pot no more, it ain't sanitary, and it's a discouragement to the plumbing business. Is that a good law?"

"For Christ's sake, Med," Greenie says, and he has a big smile on his face.

"Of course it ain't. No sense to it. But the game law does have sense to it: it says four duck and two geese a day as a limit per man. And it says you can't put corn around the blind, nor use live decoys."

"I'll tell you," says Greenie, "we could have a higher bag."

"Some years it's big, some small, depends on how many birds there are. But the law itself, now, never mind details, is it a good law?"

"You know the same as I do, Med, without some kind of a limit, you wouldn't see or hear a wing the whole season long. I guess it's a good law, but two things can spoil it: bad enforcement and bad hunters. You might say good enforcement will take care of the bad hunters. I don't know, though, when I think of some of those hunters. Talk about hogs, why a hog's a dainty feeder compared to some hunters I can think of."

Greenie glanced at Med's cap with a wicked look in his eyes. I could see Med's lips crook at the corners.

He said, "Close to home?"

"Why, yes, I could reach right out and touch one of them."

They both laughed.

"Wasn't Eddy with you that day?"

"Yes, but it was evening. We didn't start till late, still and cold, and nothing in sight, boat or man."

"What made you go over like that? You both know better. Besides, it's a good law."

Med grinned. He said, "You know how it is."

Greenie got a lost look in his eyes. He said, "It's clear, calm, late. Sun about down, but the air like glass. It's cold, too, cold enough to make your blood hum, no colder. And everything is just right. You've got a good pardner with you, you get one right, he gets another left. The dog doesn't make a sound. He stands and shivers till you say, 'Go,' real easy, and he swims out with hardly a ripple and picks up the birds quietly, and brings them back without a tooth mark on them. And they keep flying in now, a redhead, a canvasback, a goose. You're having a hell of a time, you're alive. And you can't quit, you want it to keep right on like this."

Med scratched the small of his back reflectively. He said, "I see you've been there before. But, hell, Greenie, it got so I didn't know where to put all those extra birds. I stuffed a bunch of them up under the blind, but we didn't dare leave them and pick them up later. And now it was almost dark, so we decided to take a chance on it and bring them in.

"You know that game warden, the one they call Gum Shoe? Well, we knew he was creeping around in his golightlys, had been all week. But there was not a sign of him all day, so we went ashore. We gathered all those birds up, Eddy and me; we had them under our arms, dangling in clusters from our hands, hanging on our shoulders—we must have looked like walking featherbeds. And we was quiet, didn't make a sound, not a whisper. We started up this path, and everything was still. I could even hear Eddy pant a little bit.

"And then a big, booming cough went up right behind us. Scared? We was all tensed up anyway, and that cough did it. Those damn birds began to fall. You could hear them thumping down on the path. As soon as I would reach out to grab one, another would fall, and it was the same way with Eddy. We just gave up and let them all go. Greenie, it *snowed* birds on that path."

Greenie doubled over, but he got his wind back after a while.

"Med, what did the warden say?"

Mr. Med put his hands on his hips to imitate the way the game warden stood in the path, and then said, very gently, "Have a good day, boys?"

After Greenie untied himself again, there was one thing he wanted to know.

"I expect you had some fines to pay. Was it worth it?"

"We did. And, no, it wasn't. When ducks and geese begin to cost anywhere from five to ten dollars a pound, it ain't worth it."

The Blind

A few days later, Greenie showed up again, looking for the boss. Greenie had some kind of metal contraption for recovering swallowed fishhooks, and he was aching to demonstrate it.

"He's out hauling some cruiser off Herring Island. Boat has a busted wheel."

"Right on top," Greenie said. "They never learn."

"Well, that *is* a tangle of buoys there. You've got four buoys around the island. Another one on the edge of Bennett's Point, another off Wye Town, and one on Hambleton Hills, all in the same general area, and all different colors. Besides which, the so-called island is all under water. It's confusing."

"Charts. That's what charts are for. They're cheap and they help. Just the same, it's easy to make mistakes. I ought to know. I've made a few."

"Mistakes?" I inquired innocently.

Now, Greenie has good manners. When we say that, we mean he will "set right down and talk to you." A man may be busy, he may be in a hurry, or worried, or disappointed, but he won't forget his manners.

A waterman will be working on his boat motor. The motor snorts, jumps up and down, misses, backfires. But at last he gets her tuned. She runs sweetly. The waterman shifts his cap onto the back of his head, looks up, smiles, and says, "Hear that? She sets right down and talks to you."

That was what Greenie did, even though he was disappointed in not finding the boss. He told me about a duck blind he used to shoot from in back of Tide Mill, just where Eastern Bay meets the Miles River, a lot of open water.

"She's gone now," I said. "The ice took her out about two years ago. Right on the edge of deep water in that clay. I remember that blind. Must have had eight or ten feet of water beneath her."

"She did. And the ice took her out a few days after the last time I shot from her. I took Joe with me that day, and before it was over we both wished we'd stayed in bed."

"I've heard odds and ends, but I never got the right of it."

Greenie settled down on the bench and turned the hook retriever over in his big hands.

"It was early morning and bitter cold, the kind to make your hairs stand on end, but no wind at all, a pretty day. We rowed out to the blind in a little skiff, and we waited.

"No luck. Too clear and calm, right at first, and we didn't have any decoys, just hope more than anything else.

"Meanwhile, this little breeze gets up, not a breeze, really, more like a draught, right down the middle of the bay and into the duck blind. Sneaky cold; it got into everything. We didn't pay much attention. But it kept stirring and sighing until it woke itself up, and cap'n, it began to blow. You know there's not a thing to break a breeze between Kent Island and Tide Mill, and Kent is as flat as a hogchoke."

Greenie began to rub his hands at the thought of it.

"I stood up and looked over the cedar bushes, and that blast of cold wind was enough to cut your eyeballs out. I looked at Joe. He was blue around the lips and shivering. I said, 'Joe, time to leave.' We were both willing, God knows. I looked over the back of the blind for the skiff, and there wasn't any skiff."

"What!"

Greenie nodded. "No skiff. Gone. Clean up to Deep Water Point by that time, I guess, the direction that wind was blowing, and the tide at flow, too.

"I told Joe. He looked bewildered and I don't blame him. That skiff was our only way to shore, and not a tonger anywhere. There was a boat on Tide Mill bar, but it must have left when the wind made up. I didn't even see him go. It was a lonely feeling, I'll tell you.

"Some reason or other, I'd pulled a paddle up into the duck blind when we landed from the skiff. I picked it up and dipped the blade into the water. It came up nice and new and shiny, a new coat of varnish, all ice.

"The wind was half a gale now, and the sea up with it, spray rattling like pebbles on the boards of the blind and hissing into the cedar bushes. They were soon covered with ice.

" 'Joe,' I said, 'we've got to get out of here, and quick. It's that or freeze to death.'

"But Joe said, no, let's wait for help, we can holler, we can fire the guns.

"It was no good. I knew it wouldn't be. Tide Mill is a right lonely spot anyhow, and in all that racket of wind and sea, who would hear? Even if you could hear, a gun shot, that just means some damn fool is out hunting in all that weather, it doesn't tell anyone we lost a skiff.

"We were getting colder all the time, and I knew if we didn't get out of there soon, we wouldn't be able to; we'd be too numb.

"I told Joe we'd have to swim for it or freeze, one. But Joe said, no, we'd better wait for help. I said it was only a few strokes to shoal water, we could wade the rest of it, but I couldn't make him budge. And I'll tell you, every time I looked to shore, it seemed further away. Maybe Joe was right, maybe I was wrong; but I couldn't just stand there and freeze while we figured it out. I had to move.

"I threw off every stitch I had on and jumped in."

We sat there on the bench in front of the store. The hot sun glaring up from the calm water, a moment before oppressive and eye-searing, seemed kindly now, warmly protective, the way it should always be.

I said, "I've heard people don't last long in water that cold. A few minutes to a half hour, maybe, and even then they might not pull through."

Greenie chuckled. "Honey, I wasn't in that long.

When I hit that water, I exploded: kick, claw, and scramble. Funny thing is, the wading was the worst part, it seemed so slow. But I never stopped for a second. As soon as I hit land, I threw it on, top speed all the way to the nearest farm house. I slammed right into the front door, and there was the farmer's wife looking out at me, and me stark, mother naked in a northwest gale.

"They're good folks. Now, some might stand around with their hands in their pockets, asking a lot of fool questions. Before I knew it, I was on the floor, and I began to spin like a top. Time I'd blinked my eyes once or twice, I was wrapped up head to toe in the living-room rug.

"They warmed me up and gave me some clothes to wear. Of course, I couldn't stay; I had to get Joe. I got the nearest boat with a motor in it and picked him up not long after I'd jumped out of the blind."

After a while, I said, "Greenie, I know this is not a polite question, but who tied that skiff, or tried to tie her?"

"I don't remember. I look at it this way. If I didn't tie that skiff myself, I should have. If I did tie her, I didn't make her fast. When you go out to a blind with another man, tie the skiff yourself—and make her fast."

The duck blind is simply a platform of planks nailed to stout stakes driven well into the bottom, if the blind is offshore. A rough bench is added, along with side boards to make a sort of box. To the boards branches from cedar trees are nailed so as to hide the hunters from the birds with a natural camouflage.

Most men prefer the water to the shore blind, because they believe that chances for getting shots are greater. After all, water is the natural element of their prey. But the water blind, from fifty to several hundred yards from land, can be dangerous.

Many men have been rescued from blinds by the coast guard. Others have not been so fortunate. The hunters find they are cut off by heavy seas, a swamped or runaway skiff, and lives have been lost in just this way. But where the best flights of fowl are found, that is where the hunter will go, in spite of discomfort, exposure, and the danger of accident in storm and cold.

Ducks and Geese

Turgenev in his *A Sportsman's Notebook* finds little to praise in duck hunting. He says flatly, "Wild duck present no special attraction to a real sportsman." Incidentally, when Turgenev did condescend to waste shot on ducks, he very nearly drowned because of a leaky skiff. It sounds familiar.

In the greater part of Talbot County, it is impossible to go five miles in any one direction without striking navigable water. The whole Eastern Shore of Maryland is so infiltrated with fingers, rivers, and inlets that some counties are as much water as land. The consequence is that duck and geese are the prime object of the hunter. If you take a boat ride through the oyster fleet some day late in Fall, say from Long Point in the Miles River to Tilghman's Point in Eastern Bay, every other tonger you see on the washboards will have his duck-hunting license worn right in the middle of his back.

Except by landowners and their guests, quail are seldom hunted intensively. You would have to go further south, say to the Carolinas or Georgia, to find a wide interest in quail shooting.

In addition, the local men, some of them as keen for sport as any who have ever lived, are practical as well. They want to bring some meat home with them, and a comparison between a brace of quail and a brace of duck will show you where the meat lies.

The quail hunter will point out that he has gotten his brace only because he is a first-class shot. He will also say that a great part of his pleasure is in the pretty work of well-trained bird dogs.

As for expert shooting, the quail hunter may be right, though that has been hotly debated. He is certainly right about his dogs. It is a fine thing to watch them work. But the duck hunter also has his dog, and I must say more about him later.

I can remember when the chief object, if not the only one, in shooting waterfowl, was to get something to eat. That was in the early thirties when food of any kind was frequently hard to come by. I came down to Tunis Mills from Baltimore one day in winter during the depression period, and I stepped across the lane to visit with my friends. As I went past a small, screened porch on the back of their house, I looked in. I will never forget what I saw. The top of this porch on the inside was everywhere strung with lines. On these lines hung birds in the cold air, canvasback, mallards, redheads, black duck, an occasional grey goose. They were so thick you couldn't see the porch ceiling.

My friends may have gotten tired of fish and fowl, but they surely did eat, and eat well.

But the balance has swung back the other way, now. Ducking is a sport, not a necessity, and that is what it should be. Since it is a sport, the hunter should behave accordingly and give the wildfowl a sporting chance.

I will tell you what I saw one day while tonging on Long Point Bar. Right beyond the point, a creek makes in called Hunting Creek. On the other side of the point, lies the Miles River. On both sides, in the creek and on the river, are many duck blinds. The occupants of these blinds were after geese. Now, geese soon grow wary as the gunning season progresses. In the daytime, they tend to gather in large flocks on the open water, well away from shore and, hence, away from blinds. Myself, I think the geese are to be admired for showing good sense.

But the hunters played it dirty. They sent out one of their number in a fast boat which roared up and down Hunting Creek and the Miles River and drove up the geese wherever they were resting. The boat kept putting them up, never giving them a chance to light. They had to fly somewhere. Naturally, numbers of them flew past the blinds and were killed. One of the geese staggered, picked up his wing stride again, and flew right past me. His side was streaming with blood.

Duck and geese are often hard to kill. The cripples seldom last long, however. They die of their wounds, or hawks and eagles pick them up. And if the hawks don't get them, the tongers will. They run down the cripples and kill them. I have seen many a culling board with a brace of duck or a goose on it. I regret to say that, on the whole, the oysterman looks with a kindly eye on your Saturday hunters in the rented blinds. They produce an excellent crop of cripples.

Deer

Hunting habits have changed radically in my time. For example, the introduction of deer into some of the

Eastern Shore counties, including Talbot, has put what amounts to an entirely new animal in the field. For over twenty-five years I had never heard of deer in Talbot, nor had anyone else. They were merely a legend, an almost forgotten memory, like the otters which once inhabited the headwaters of the Miles River. In the last five years, deer have become solid flesh and bone. Dozens are killed every season in Talbot alone. They thrive in this mixture of woodland, open farmland, brush, and river bank. There is plenty of forage, and the climate, compared to that of New England, is mild.

For many winters, I lived in upper New York State. In that countryside, deer had been known for generations. They were a commonplace. The consequence is that native car drivers paid attention to metal signs which read, "Deer Crossing." They knew, without thinking, that these signs mark places where deer commonly cross the highway.

In Talbot these signs have not yet taken hold of the popular consciousness. Every year, cars are smashed into junk. A deer is a right hefty animal. When a car hits one at high speed, the results are usually devastating.

The experience of two oystermen will demonstrate just how little the deer is a part of our traditional hunting knowledge. I will invent a name for this pair of tongers because, while they made a gallant fight, they acted from the most profound ignorance. It is always annoying to be told, "You just don't know."

Well then, George and Joe Harris put out of St. Michaels early one brisk October morning to go oystering. That is the way they made their living. They knew a lot about oystering, but almost nothing about deer, except that they are said to be very good to eat. It is unlikely that either had ever seen a deer, except on TV or a hardware calendar. But now they were about to meet one in the flesh. Even though these animals are famous for swimming ability, I admit that to find one in Eastern Bay is far from likely. Nevertheless, that is where the meeting took place, away out, as the Harrises later said, "in the middle of nowhere."

They circled the swimming animal a couple of times, and came to the conclusion it was a deer. Unlike a cow, this thing had a long neck and crooked horns. It looked like a deer.

"Lot of meat there," George said, kicking the tiller over to come alongside.

"Yeah," said Joe.

"How do we get him?" said George.

It was a good question. A pair of twenty-foot oyster rakes is not admirably adapted either to the killing or the capture of deer. You can't do much with a culling hammer, either. The head weighs one-quarter of a pound. There was nothing else.

But these were fairly rugged men accustomed to doing things for themselves.

George said, "Tell you. You get him by the crotch. I'll grab him around the neck. We both heave, and up he comes."

"Yeah," said Joe.

They did that, and up he came, too.

A deer is a soft looking, rubbery seeming, bouncing kind of a beast on thin, stiff legs. But anyone who knows will tell you that every inch of him is constructed of high-grade spring steel. And those little blocks at the ends of his legs are called hooves, but they happen to be small butcher knives.

If anyone had asked me whether two men in a thirty-foot boat could wrestle down a deer and hold him long enough to truss him with the anchor line, I would have said, "No. No, sir."

These men did it, and any waterman in St. Michaels will be proud to tell you so.

When they got through spitting out hair and wiping it from their eyes, they looked at one another. Their jackets were gone, torn to shreds. Shirts fluttered in rags from their belts. They were naked to the waist, battered, bruised, but unbowed.

They had a lot of meat, but it was still very much alive. They decided to take it back to St. Michaels.

You know how news travels. There are eyes everywhere, even in the middle of Eastern Bay, and an epic struggle of that kind would not go unmarked. So the Harrises had an audience waiting for them when they made fast at the landing. Unfortunately, among those present was a game warden, who at once relieved them of all that meat. With the help of bystanders, he packed it neatly, though still kicking, in the trunk of his car. He drove the car outside of the limits of the town of St. Michaels and there released the deer, which was last seen bounding off in the general direction of Pea Neck.

The Harrises now know a great deal about deer, more than most of us will ever learn through direct experience.

Small Game

Before turning to small creatures without wings, I should mention the dove, which is still a highly esteemed bird in Talbot, as it is all over the southern tidewater. Ever since I can remember, I have heard that the dove is disappearing. Nevertheless, he is still with us. The official bag limit per day per man stands at twelve this season of 1960. And a dozen dove will make a respectable pie.

At one time, coons and muskrats were hunted intensively for their pelts, but that is no longer the case in our vicinity. There is relatively little market for the furs. In the twenties, Russia and other European countries were importing pelts in large quantities, and a good local price was easily obtained. This market is either no longer available, or it has ceased to exist.

Then, too, fashions have changed. The disappearance of the raccoon overcoat is a well-known example. In general, coarse furs are not so extensively worn as they used to be. It is said, also, that the increased production of synthetic fabrics has hurt the fur trade.

For many years, some of my neighbors ran a few trap lines on the side, after work, before work, on the weekends. They picked up some extra income. But now the price offered for both coon and rat pelts is so low the extra work of shooting and trapping is not worth the labor. Nowadays, the nearest place where muskrats are still trapped intensively for market is in the big marshes of the Choptank River.

There is a limited local market for the meats of muskrats, which used to be thrown to the dogs. But, again, the price is low. The market for meats is not likely to enlarge, even though the muskrat makes excellent eating, like most rodents. When he is skinned, however, he looks exactly like a large barn rat.

Coon hounds were a fairly common sight at one time, but they are seldom met with any more. The reason is that coon hunting, which was once a traditional sport, has now fallen into almost complete disuse.

One man in the Miles River Neck, Melvin Johnson, still continues to hunt coons as a sport, almost a hobby. He is a real enthusiast. He hunts, of course, at night, since the coon is a nocturnal animal. He has a small trailer containing individual compartments for his hounds, with the name of each dog printed in bold letters over its particular cubby hole. In this way, he can take his pack to any desired location by hitching the dog trailer to his car.

For a long time, now, coons have not been very numerous in the United States, and those who know claim that this animal is rapidly disappearing over most of its range, which is everywhere in the nation, except the Northwest and the drier parts of the West. Furthermore, it is calculated that two hundred acres of woodland are necessary for the continued, year-to-year yield of one adult coon.

Most of the Eastern Shore of Maryland is still well endowed with woodland. But another factor enters the calculation of yearly yield. There are literally hundreds of miles of shoreline in countless creeks, coves, bights, sounds and rivers. I am perfectly certain that fewer than five or six acres of shoreline are capable of sustaining the same yield as two hundred acres of woodland, a figure that seems high to me in any case. A narrow but continuous strip of thicket, brambles, trees, myrtle, duck grass, and marsh occurs between plowed land and low tide. It contains an abundance of food in all seasons but winter. The coon solves that problem by hibernating.

In this fertile strip, all kinds of insects abound, and the coon likes a fat grub or a meaty locust. He is not above doing a little fishing either, for which the bank position is obviously favorable. He will eat snakes as well, and any number of water snakes swim in the shoals or take cover in duck grass.

Wild cherries, persimmons, and berries flourish in this band. For a further balance of diet, the landward side of the strip is often planted in the food he likes best—corn on the ear. At the first approach of a farmer, escape and concealment are not twenty feet away.

If you add to these advantages the fact that coons are probably the most intelligent wild animals in America, it is difficult to see why any normal coon would need more than a few acres of shore front property in order to retire in comfort for the rest of his life.

I have seen many woods coons and bank coons, and I agree with local opinion that the bank coons are bigger. The difference must be due to a better diet. So coons hereabout are fat, saucy, and abundant even when their numbers are thinned by shooting, and they are now seldom shot. The result is that they have become a nuisance.

In the first place, he sees no reason for not eating your garden corn just the night before you have decided to pick the first juicy ears. Somehow, the coon is always one move ahead of most gardeners, including me. I am certain that I am not as smart as a coon, and never will be.

In the second place, coons have a special place in their scale of values for hens. They can resist a fat fowl only a certain length of time; then they have to eat that bird if it is at all feasible. Many a villager and farmer has wakened in the night to the agonized squawking of a hen, only to find her plucked from the roost, and both thief and fowl vanished. One of my neighbors witnessed the abduction of a hen by full moonlight. The coon had climbed a high wire fence, seized the hen, and was climbing out with her when he appeared on the scene. Unfortunately, he had forgotten to bring a shell for his shotgun. The coon coolly leaped down from the fence and disappeared with the hen.

Just the same, he had gotten a warning. He had no desire to be caught inside that fence again. After that he did what I can only describe as making passes at the chickens. Any unwary hen, who, for coolness, decided to dig out a resting place in the ground near the fence, was liable to lose its tail feathers and what little mind it possessed. The coon would grab at the hen through the wire from outside. Several of my neighbor's hens had naked tails. The only satisfaction the coon could possibly have derived from such attacks must have been purely sporting. Tail feathers are not edible.

These animals have as strong a desire for comfortable shelter as men have. The result is that they will often move into a house as soon as men move out, and sometimes before.

Several unoccupied houses in Tunis Mills have been inhabited by raccoons who paid neither rent nor taxes. They live, not in the rooms, but in the spaces between weatherboard and plaster, floor and the ceiling of the room below. They will get in any way they can, but, usually, they enter these dens by going under the house, since most of the village homes are not built on foundations, but are raised a little above the earth in order to allow circulation of air. We have no cellars. The water table is only six or eight feet below the surface of the ground. When open wells were common here, it was a familiar fact that the water level rose and fell with the tide in nearby Leeds Creek.

Raccoons can cause extensive damage in a house. A home near ours was empty for several years. As I walked by it one morning, I heard a curious scuffling, thumping noise in the space between the shingle roofing and the tongue and groove ceiling of the front porch. Coons had moved in. I at once wrote a note to Howard, who came down occasionally from Kent Island in order to look after his mother's house and cut the grass. A few days later he met me in the village and we opened the house. The coons had already done considerable damage to the front room ceiling. A careful examination of the outside of the house showed their method of entry. The bark on a large catalpa tree was slightly scuffed. The coons climbed the tree, ran along a big limb overhanging a side porch, climbed the weatherboarding to the eaves, and entered through a separation in the cornice.

Howard said, "I'm going to catch one coon anyway. If I get one, the other will leave."

I said, "Good. How?"

"Easy. You know what a wire noose is? You make a noose out of thin, strong wire and set it over a hole that barn rats are using. Rat comes along, sticks his head in the noose, jumps forward to get free, fights the noose until he strangles himself. I've caught many a rat that way. All I have to do is fasten my noose to a strong nail and arrange it over that hole in the daytime. When the first coon comes out at night, he's going to hang himself. I'll put the noose up today. You come over tomorrow, will you, and take him down?"

"Sure," I said. "I'll take Mr. Coon from the noose, if he's in it. But he won't be in it. A coon is not a barn rat."

Howard put up his noose and drove back to Kent Island in perfect confidence. It was a good noose so placed that the coon had to thrust his head in it to get out.

The next morning, I went over to take down a hanged

coon. The noose was empty. What we tend to forget is that a coon has hands. The first coon out the night before had noticed the wire noose. With his clever hands he had bent the wire neatly out of the way, and so disposed of all danger to him and his mate. I wrote the sad news to Howard.

He returned with a gleam in his eye and a can of red pepper. The theory is that you put the pepper on the route used by the coon. He steps in it. Then, as he would do with any foreign matter, he licks it from his paws. When he does that, he gets a hot mouth he will never forget, and he leaves the area for good. That is the theory.

Our coons simply stopped using their customary way of entry and found a less convenient one. After the first heavy shower, they returned at once to their usual route by way of the catalpa tree, the red pepper having been washed away by the rain. I had to write another note.

We then tried trapping them in a highly-praised and widely-advertised contraption consisting of a large, stout wire box having an entry at either end by way of suspended doors. On the floor of the box was a very sensitive treadle and a container in which to place bait. The idea was that the coon would go into the box for the bait, step on the treadle, which in turn slammed down both doors, and there is your coon, neatly trapped. We tried all kinds of bait, including canned salmon, a well-known coon delicacy.

I must say, the trap worked very well indeed. On one night, we caught a big, fat slug, thus demonstrating the extreme sensitiveness of the treadle. On another, we captured a large, striped tomcat. But never a coon.

Once in a house, coons are the very devil to oust. These animals denned in the house for a year or more, rent-free, tax-free.

Mr. George, who used to live next door to me, had his house invaded early one March by a bitch coon. He did all he could to frighten her away by pounding on the walls and rapping the ceiling with a broom handle, without any success at all. To make matters worse, the bitch was in heat. Every night for a week, she was visited by two bucks who fought over her and chased her from sundown to sunup. Neither Mr. George nor his son got much in the way of sleep. Sixty-odd pounds of fighting, leaping, scrambling, squalling coons in the ceiling can really make their presence known.

Finally, however, the lady made her choice and settled down with her mate to the coon's usual monogamous existence.

A few weeks later, Mr. George's son returned one rainy day from work, ate supper with his father, then went to the front room where a lamp on a marble-top bureau threw a good light for reading the newspaper. He was annoyed by drops of water falling on the papers spread over his lap. He called his father and told him the roof must be leaking, pointing out a wet, discolored patch of plaster directly over the chair and lamp. Mr. George sniffed the air.

He said, "That's not rain. That's pure coon water. Can't you smell it?"

And that, of course, is one reason why coons do so much damage in a house. It is not for nothing that one of the coves on Leeds Creek is named Coon Hollow.

I have gotten several informal visits from coons, especially during thunderstorms at night. I told you they like comfort. I know all about the mess a coon can make and how he gets into the walls. I also know that another neighbor of mine hustled into the kitchen one summer's morning, gathered some kindling for her old-fashioned cookstove, and lifted one of the lids to build a fire. A coon jumped out of the stove. It is still not certain which of the two got out of the kitchen door and into the back yard first. The coon had come down the chimney.

I have chicken wire all around the bottom of my house, and I have the same over my chimney. I have a good deal of admiration for raccoons, but I am not going to live with them if I can help it.

Other small game, such as squirrels and rabbits, are regularly hunted in season, here as elsewhere. The skunk has his odor yet, while his pelt has fallen in value, so he is no longer trapped. He is now merely an occasional nuisance, except to one man, old Mr. Burroughs.

In his late eighties and an expert boat and ship calker, Mr. Burroughs is as pleasant and as likable a man as you could wish to meet. But he bears an inveterate hatred for all skunks. Those who know of Mr. Burroughs' ceaseless warfare often invite him to kill a skunk which has become a pest, and the old gentleman never disappoints them. He loads his shotgun and jumps in his car. I sometimes wish he had a brother with an equal hatred for raccoons.

Chesapeake Bay Retriever

It would be unthinkable to leave the subject of hunting without paying tribute to the Chesapeake retriever, the dog which is used almost exclusively for all duck hunting. He is a large, well-made, even-tempered, and intelligent animal, the most characteristic sporting dog of the Eastern Shore of Maryland, for he was first bred in Talbot County. In fact, he is one of the two sporting dogs of American origin, the other being the brown water spaniel.

To most dogs, the water means a queer but very entertaining place along the shores where all sorts of exciting things happen. To a Chesapeake, water is a second element. His swimming ability is so famous it has given rise to a legend still related in the Bay region. It is said that he is the result of a cross between a female dog and a dog otter. A more prosaic account is now generally accepted.

At the beginning of the nineteenth century, a Newfoundland vessel sailed for England. Severe weather very nearly sank her, but she was able to draw the attention of an American ship which rescued the crew and two puppies. These youngsters were of the Newfoundland breed of big dogs with short, very heavy hair. Both crewmen and puppies were landed in Norfolk. One of the puppies, the male, was sent to Talbot County and was there bred with hound bitches. The progeny were in turn bred with still other hound bitches. At some point in the generation, it is now agreed, the dogs were then bred with water spaniels. What emerged was a dominant line of blood which ran very true; in other words, a genuine breed. It is the Chesapeake retriever.

The breed is subject to color change, or rather a shifting of shade, but this characteristic can be used to great advantage. Since they breed every shade from the light straw color of winter duck grass to a brown so deep it is almost black, the hunter can choose whatever shade most easily blends with the natural color of the hunting area.

The Chesapeake weighs about one hundred pounds, and he is very strong and durable. He will swim long distances to retrieve and return with a big grey goose without apparent laboring. His pursuit of a hurt and diving duck is absolutely relentless. When the duck is down, the dog simply waits. He treads water and he keeps his place, regardless of cold, sea, or tide. As soon as the duck bobs up, he is after it again. He is a dog of great heart and strong will.

I have even known Chesapeakes to dive after ducks and chase them to the surface, though this is not a usual practice. I don't suppose diving is or should be encouraged, since it is exhausting work for the dog, and the same result is attained by treading and waiting.

I remember a big, beautiful Chesapeake that had a way of amusing himself with a flock of ducks belonging to one of my neighbors. When the dog ran up on them, they took to the water in a mob. They had their wings clipped, so their only means of escape from the dog was by diving. He promptly followed them under. There was little depth, no more than six or eight feet. He really had his fun with those harassed birds. When he had chased them as far as honor and reputation demanded, he would swim back, panting heavily, but stay offshore in the hope that someone might throw an oyster shell. Those he could not catch before they struck the water, he dived for and often retrieved.

They are excellent dogs to take along in boats, because they are happy on water, never frightened or ill at ease, as so many dogs are. I know one oysterman who used to have as his partner all winter long a dark brown Chesapeake. The tonger was sometimes cold on bitter days, but not the dog. His double suit of hair, the undercoat of which is very short, oily, and thick, protected him from all weathers and all waters. He was very much interested in the work, for he carefully inspected every rakeful of oysters dumped on the culling board. The tonger often complained that his dog no longer culled oysters for him the way he used to do.

A Chesapeake will go into anything. Unlike the golden retriever, perhaps the most beautiful sporting dog in the world, he never refuses. He will shove floating ice aside with his shoulders or scramble over it to retrieve a duck. Since he is a powerful and confident swimmer, no sea or tidal current frightens him.

Many people have noticed and recorded the Chesapeake's behavior in the presence of other dogs. He is not a noisy dog. Bark and bluff are not his style at all. He tends to keep to himself. If another dog worries him beyond endurance, the Chesapeake makes a sudden irresistible charge and bowls over any opposition with his powerful chest and shoulders. He then stands as if carved from stone and delivers a very serious warning in the form of a deep-chested and continuous growl. Having delivered his message and seen that it was comprehended, he then walks away with superb indifference.

Obviously, these are not ideal house dogs or lap pets, as the cocker spaniel is, for example. The Chesapeake is patient, aloof, tolerant, and good-natured. But he does not fall all over himself to attract attention or favors, and he does not expect you to fall below his own high standards of social behavior. You can talk to him and scratch him behind the ears, and he will appreciate the intention, rather than the attention. But don't overdo it. If you make too much fuss over him, he will walk away. If you keep it up, you may hear that deep-chested growl I have mentioned; at which point it is best for you and the Chesapeake to part without too many protestations of regard and esteem.

And so we leave him.

boats

7

Wheat

When we were boys, my cousin Bob and I always did one thing first when we woke each morning of the wheat harvest season. We went to the west window of the bedroom and looked toward the wooden bridge spanning Leeds Creek. Often, we had our reward. There at the landing built out from the bridge lay the sailing vessels. Rising from the morning mist, as slim and clean as a wand, was the single, raking mast of a skipjack, or the twin thrust of a fore and mainmast in a bugeye, or the stout, sturdy, dependable sticks of a big-masted schooner. It was like magic.

After snatching a quick breakfast, we ran as fast as we could to the bridge and engaged the skipper in a short conversation which generally ended with permission to come aboard. In those days when boys hung around a job, they were quickly put to work. So we helped the deck hand fold and stow tarpaulins neatly on the foredeck, because they would be needed later to cover hatches or to put over grain sacks left on top of the deck when the hold was full. A scale was then set up on the landing, and a wooden chute was unlashed and placed with one end on the top of the landing, the other just inside the hatch, and made fast. By this time the first wagons drawn by mules came rattling over the loose, wooden planks of the bridge. When the wagon was backed into place with its tailboard next to the scale, the first bag of wheat was handed down and weighed. The farmer whose wheat was being shipped stood with a notebook and pencil by the scale and made an entry of the weight of the bagged wheat, and the captain of the vessel did likewise. At the end of the lading, the figures were added up and compared. As soon as the bags were weighed, they were quickly opened by pulling out the light cotton thread holding the mouth together, and then emptied into the chute. The hissing sound of grain sliding on the chute and the flash of sunlight on this golden waterfall continued all day and the next, too, if necessary, until the vessel was laden. All was then made shipshape, the hatches secured and covered, and the extra bags which could not be poured into the crowded hold were protected by tarps lashed in place. At the top of the next flood, the vessel was eased from her mooring, and a yawl boat with a kicker engine was lowered aft and secured in a bridle. She would push the vessel out of the confined waters of the creek and into the Miles River, where sail was set for the trip to Baltimore and its granaries.

Where there was no convenient landing like the one at Tunis Mills, skipjacks, bugeyes, and schooners frequently poked their stems into the shores of the remotest farming lands, wherever the water was navigable. Before

16 LOADING WHEAT TO BE BARGED TO WHEAT SCHOONER

17 WHEAT SCHOONER

18 OYSTER DREDGING BUGEYE

trucking became so widespread, these vessels performed an important and cheap service.

Since the ship captains went so often ashore in remote places to arrange with farmers for the shipment of wheat, they were frequently bothered by biting dogs who saw a complete stranger afoot where none was expected. Vicious dogs became an occupational hazard for ship captains.

Captain Will Roe, who was master of a sailing vessel for many years, decided one day that the time had come to protect himself. He intended to live a little longer yet, and he wanted the full use of his limbs. So, the first chance he got, he secured a fine straight piece of green hickory, and on the end of it he fastened a heavy brass knob. It looked for all the world like a walking stick, but it was not. It was a dog stick.

One fine summer's morning, he got in his yawl boat and went ashore with the intention of letting a farmer know that he and his vessel were there and both were ready to load wheat. As he approached the front porch of the farmhouse, Captain Will saw the farmer's wife, rocking and sewing. Near her feet was a big dog. Within speaking distance, Captain Will lifted his hat and bade her a good morning. As soon as Captain Will raised his hat, that dog got up and made for him without a sound. Letting his hat fall back into place by the force of gravity, Captain Will rapidly reversed his stick so as to grip the wooden end and expose the heavy metal knob. He was just in time. The knob met the dog as he was in the air and headed straight for Captain Will's throat.

It was a shrewd blow. The dog sat down on his haunches and studied the heavens. He then got up and began to walk slowly in circles.

The farmer's wife said, "I never seen him do that before."

Again, Captain Will raised his hat. "No, madam," he said, "I expect not. He was never hit that hard before."

As boys, we had no outboard motors on our skiffs. Often the captains of sailing vessels would give us a lift when they left Tunis Mills. After the ship reached the mouth of the creek, our skiff was cast off, and we would wade for miles on the Second Point flats looking for crabs. The three-or-four-mile row back to the village seemed easy after having had a lift one way. The captain, before casting us off at the creek mouth, would invite us jokingly to "work our way" to Baltimore. I am sorry now that I never had the chance to do that, but these craft were already thoroughly familiar to us. We could not imagine a summer without them.

But that summer came. With one or two exceptions, sail is no longer a matter of cargo business on the Bay. Grain, heaped golden in the hold and carried from Talbot and the remotest reaches of the Bay by a force as natural as the wheat itself, is little more than a memory. But the magic of those masts, come in the night while the village slept, and leaping strongly from the mist of morning, will always remain a part of me.

There is one vessel in particular that stands out clearly in my memory, all but her name, which I have forgotten. She may have been the *Florence* or the *Emily*. I will call her the *Florence*, since she has to be called something, but I make no pretence to accuracy in her name. She was a big two-master, and, as I recall, schooner-rigged, and she came every June to the mouth of Leeds Creek to carry off wheat from a farm whose shores faced both the creek and the Miles River. The river side was very shoal, but there was an indentation or hole on the creek side close to shore with good water in it. She would get in as close as she could anchor and not ground on the ebb; she often did anyhow, when her holds were crammed with wheat and the bags were piled high on her decks. I often thought the peculiar sheer line of the *Florence* was the result of having squatted down on so many shoals and mud flats. The line from bow and stern rose slightly to midships, so that she seemed about to break her back. When laden, her captain would have to wait for a high tide before she would budge an inch, even under sail and with the added help of her yawl boat. A big negro deck hand would hold his own shirt aloft to catch the air in a kind of supplication, a prayer to wind and tide.

In this case, the grain wagons were driven right out into the shoal water up to the wheel hubs, and the grain bags were passed from the wagon beds to a barge. By the end of a day's work, the backs of both crew and farm hands must have been in much the same precarious condition as that of the *Florence*. Every grain of wheat was poled laboriously from wagon to ship. The bags were lifted by boom and the muscles of the men straining at the rope's end. The wheat was then emptied into the hold, where a crewman shoveled the cargo to ensure proper stowage. I have done a little wheat stowage of that kind, so I know what it is like.

In the first place, there is very little footing in loose grain. Every step taken must be made ten or fifteen times to one stride in normal conditions. It is a most annoying and frustrating circumstance to wade furiously toward a bulkhead no more than a couple of yards away and not get there, not for a long time. And you work in overpowering heat, the body at any given moment as much water as flesh. The air is a cloud of dust and palpable, floating chaff, from which must be extracted, somehow, enough oxygen to keep the shovel swinging and the feet in their endless yards of a journey which, after all, may take them no further than the next bulkhead in a half-hour's time. Skin is soon creamed with a mucky, itching dust that gives no rest; eyes burn in red-rimmed sockets.

The good old days, hmm? It is a wonder after fourteen, sixteen hours of loading the wagons, loading the barge, poling the barge, unloading the barge, shoveling the wheat; it is a wonder, I say, that a man among them had any back at all. And yet, when the job was done, though they swore, grouched, groaned, and flexed their driven bodies, still, in some corner of the mind lurked a feeling of gloomy pride, perhaps summed up in a phrase such as this:

"Well, I guess we showed them how to stow wheat today."

Showed whom? Why, no one in particular; except themselves and a vague, mocking, sneering spirit that intimates he has seen schooners loaded in *half* the time by better men than *they* could ever think of being. It is an obscure pride, perhaps, but a very real one. The man who touches a lever and so causes a flexible pipe to nuzzle freight car after freight car and suck up a whole train load of wheat in a matter of minutes may find the work easy, even agreeable. But he has not shown "them" anything.

However, what I remember most vividly about this vessel is the way she died, for boats die just as men do. It can not be said, though, that her end was dignified.

Her captain ran her into St. Michaels on the edge of a hurricane, snugged her down, and left her, he and his crew, to spend a welcome night ashore. She was perfectly safe, well away from all traffic; and, anyhow, there would be no traffic, not in a building hurricane.

Now, a hurricane is not like a squall. A squall strikes like a viper, quickly, viciously, and without much warning. A hurricane takes its time. For twelve hours or so, it leans a great shoulder on the coast, puts more and more weight against the water. The wind rises, deliberately, four or five miles an hour more every hour, until it screams constantly, without let-up, without remorse.

By midnight of the afternoon the schooner dropped her anchor in the safe harbor of St. Michaels, the wind was a steady unrelenting seventy miles an hour, with gusts measured to eighty-five. For hour after brutal hour, that great shoulder leaned heavily from the southeast and crushed a large portion of Hatteras through the capes and up the Bay.

Then the wind backed in a gradual arc to northeast and had its fun with the now exposed *Florence*. Under such pressure she did the only thing a ship can do. She chewed her cable wildly until it parted. She ran away, crippled back and all, and it is hard to blame her. What we had on the Shore by morning can best be described as hell and high water.

A few days later, I went down to St. Michaels to have a last look at the old lady. The usual tide marks were showing again by that time. There she lay, right in the middle of a flower garden, high and dry. A few dahlias had responded to the drying sun and now splashed brilliant colors on her somber hull. She was canted, of course, her masts pitching at a crazy angle. She was at rest, her keel hard on the last ground it would ever touch. Not even the magic supplication of the negro deck hand and his lifted shirt would ever make her sail again.

Log Canoe

Any discussion of craft native to the Chesapeake Bay must include some mention of the log canoe. This boat originated in modifications of the Indian dugout canoes, and, like them, was made of log. Instead, however, of using a single large log, the Chesapeake canoe was built with from four to seven separate logs.

A typical canoe was built completely from start to finish by the builder. He did not order his materials and then get to work. He went out into the woods and searched for his materials among the standing trees. He looked for trees of the proper size, shape, and quality.

He selected first a tree from which he could cut the center log for the canoe. This keel log had to be cut from the biggest and straightest tree the builder could find, which he then felled himself and trimmed of branches. He was careful to save all of the curved crotches where the limbs entered the bole of the tree, because he would need these pieces later to form the knees of the finished canoe.

The keel piece was then squared and shaped by broad-axe in such a way as to begin the deadrise of the bottom and the inboard upward curve on either side of the keel. One end of the log was roughly pointed to approximate the cutwater at the bow.

The boatwright then went back to the woods to find two trees suitable for making logs that would be placed one on each side of the keel log. An angle was hewn on one side of each log where it abutted the keel log in order to continue the deadrise of the bottom and to approximate the round bilge on the inside. These were called the "garboard logs."

Two more logs, the "wing logs," were now needed, the most difficult of all to find. They had to be selected from trees having a natural curve in them. When the trees were finally selected, the logs cut from them had to be of exactly the natural curve to fit the desired breadth of the hull. At the same time, these wing logs had to be of such a size and shape that the inner side could be trimmed off square to abut upon the garboard logs.

So the builder had to be an expert timberman as well as a boat builder.

The logs were then temporarily spiked together while the whole assembly was hewed into the rough shape of what would finally be a sleek, round-bottomed hull which came to a point at both the bow and the stern.

The whole assembly was then taken apart, and each log hewed out singly to form the inside of the boat. After cutting a slot in the keel log for the centerboard, all of the logs, now shaped to finished size, were reassembled.

The problem of making a perfect fit or join between the logs must have been a major one. The logs were pressed tightly one against another along the entire length of the seam. The builder then started with a saw at one end of the joint and sawed through to the other end, thus cutting away all rough spots on both logs along the seam. After several sawings of this kind, the result was an excellent joint.

All of the logs, now perfectly matched to one another, were then joined together, with either iron or wooden pins. Much expert work still remained to be done. For example, the sheerline had to be carefully worked and fitted and the trim completed. All in all, an enormous amount of skilled and painstaking labor went into the building of a log canoe.

Anyone with a genuine technical interest in boat construction and design should consult an excellent monograph by M. V. Brewington, entitled *Chesapeake Bay Log Canoes*. It is the only source of detailed, step-by-step information that I know of.

These boats were made for work, not for pleasure, and they were strong, able, and fast. Some of the larger canoes were used for short-haul cargo, but the vast majority were employed by watermen for commercial fishing, including oystering. There is no doubt whatever that log canoes were beautifully adapted to the working demands of net fishing and tonging. For one thing, their shallow draft made them ideal for use in the Bay, which is predominantly a shoal body of water. Remember also that most gill netting and tonging work is carried out in waters ranging from two to fifteen feet in depth.

So completely fitted were these native craft to the needs of watermen that in 1880 there were 6,300 canoes in use on the Chesapeake, and boat builders were turning out 175 of them every year. Captain Robert Lambdin of St. Michaels built a total of 68 canoes between the years 1865 and 1894. During the same period, there were probably two hundred or more canoe builders in the Bay area.

The sail rig itself was unique, consisting of a sprit or spar and a club. The sprit ran from a point much higher on the mast than in an ordinary boom sail to a club lashed to the after or trailing edge of the sail. The tapered or club end of the sprit was fitted into a hole bored in the middle of the club, and was held at the mast end by a lashing fitted in a notch cut into the sprit.

I have sailed this rig, though not in a canoe, and it must have been perfect for an oyster boat. The rig draws powerfully, spills little wind, sails almost in the eye of a breeze, and can be struck and furled with little trouble, simply by loosening the lashed sprit at the mast. The sprit can then be lifted from its hole in the club and set aside out of the way. The sail is furled by rolling it on the club to the mast and throwing a couple of half hitches around both sail and mast. There is no boom in the way, which is important for an oysterman, who must have his hull free for culling board and the stowage of oysters. There is no space for booms, plus the normal working and stowage, when not in use, of several pairs of rakes, ranging from sixteen to thirty feet in length.

To sum up, then, log canoes were shallow draft, centerboard, had little beam in proportion to length, sailed like the wind, and stepped a mast fore and aft, thus leaving the middle of the boat free for working.

I can tell you one thing, though. If, as a matter of curiosity, you should ever be tempted to try a club and sprit rig, watch the club when making sail or furling. The club is normally made of dogwood, one of the heaviest and hardest of woods. In a stiff breeze of wind with the sprit not holding the club in place, the club is a club indeed. Loose at the end of a snapping sail, it would think nothing of knocking your brains out.

The log canoemen knew their boats and their rig. No man will ever learn to handle them better, for they are no longer worked, and never will be again. The men who could build them are dead, and the old canoes are now fitted with auto engines or lie rotting in the mud at the heads of coves and inlets.

A few log canoes are still raced once or twice a year on the Miles River, and a pretty sight they are; but racing a boat and working her are different matters entirely. Anyone who has tried to troll under sail will know what I mean. Not that the oyster fleet never raced. It did, almost daily. The men quit work in the winter afternoons whenever they had made what they considered to be a satisfactory catch. The consequence was that three or four log canoes might be leaving a bar at much the same time. Naturally, they raced to port. They had fast boats, and each man was proud of his craft and its special virtues, while, privately, each cursed the exasperating faults and idiosyncracies every boat possesses. Individual invitations to race were made so often that the challenge was reduced to a well-understood formula. If you encountered another canoe on the oyster grounds and you wanted to race her home after work, you pointed your finger at the owner and called, "I want you!" If the owner nodded, it was understood that, after the last rakeful of oysters was culled for the day, the race was on.

But the really exciting occasions arose when a sudden and violent change in the weather caused a general exodus. On those days, Eastern Bay and the Miles River were flecked with sail as far as the eye could see. Then the whole fleet raced with all its heart and all of the sail it could possibly carry and not founder.

Old men, most of them dead now, who were young when they worked in canoes, have often told me of an especially disastrous, sudden, and violent change in the weather, the kind for which the Chesapeake Bay is still noted.

That morning showed a dirty sky the color of old tripe and a sea with a long, slick swell. It was said later that many of the watermen distrusted the look and feel of that day, but the fleet had been ice-bound the week previous and it was hungry for oysters. There was a southeast draft, enough to sail them far from port and to the rich bars of Eastern Bay, Caffee, Turtle Back, Herring Island, and Bennett's Point. All morning, and for most of the afternoon, the sky and water remained unchanged, ominous and uneasy. The fleet tonged, the culling boards were filled and emptied, the heap of oysters mounted in the bottoms. At three o'clock, a breeze rippled the water and rocked the anchored fleet at the same instant. From Hambleton Hills to Bennett's Point, shore to shore and over a distance of several miles, sails sprang quickly to life as the fleet, with one impulse, put on wings. Even a landsman, looking to westward over the Bay, would have been disturbed. A huge bank of inky clouds sponged out sky and water. Behind it, nothing could be seen. It swallowed Kent Island with a rush. Beyond Tilghman's Point, it rolled over a bugeye dredging for oysters and blotted it out of existence. It advanced on Tilghman's Point and swamped it. The fleet sailed.

As it sailed, it raced. Few of the watermen were willing to set less than all sail. They were in a hurry to beat the weather, and, while storms were bad, sharp tongues at port for last in were sometimes worse. There was no thought at the time of jettisoned cargo, though almost all of the canoes were heavily laden. A man who leaves oysters where he found them might as well try another trade. Later, of course, many would have been willing to leave the oysters on the bottom, or even try another trade, but by that time it was too late. There were men on that day who never saw port again. They had raced the fleet for the last time.

I suppose everyone regrets the loss through obsolescence of a fine piece of equipment, beautiful in itself, and efficient and workmanlike in its daily use. Still, it is possible to take a too romantic attitude in such cases. Brewington ends his monograph on the Chesapeake Bay log canoe as follows:

> Where once the waters of the Bay were alive with hundreds of trim canoes, all busy sailing to and from their work, one now finds only oily, grimy motor boats on the oyster beds.

I heartily sympathize with his point of view. Nevertheless, it must be admitted that, on the purely practical considerations of comfort, safety, and ease of work, the best thing that ever happened to the oyster fleet was the introduction of the gasoline engine.

I have talked many times to Charles Draper and Charles Banning, a pair of Tunis Mills tongers who worked for years in log canoes. Captain Draper is now dead, but Captain Banning still remembers how extremely difficult and trying canoes could be.

In the season, these men woke before dawn and walked several miles to their craft moored in Lloyd's Creek. There was no other means of getting there.

Very often in the Bay, mornings are completely calm. These men could not afford to wait around for a wind, however, so they rowed their canoe with sweeps. It is hard work to row a heavy vessel not designed for that purpose, but, if no breeze met them at the mouth of Lloyd's Creek, they continued to row, sometimes as much as three or four miles, until they were on the ground where they intended to tong.

In the first three months of the tonging season, it frequently happens that no wind blows the day long, or the air is light and fitful, from every point of the compass. So it happened that these two oystermen sometimes rowed back as well as out, and that on top of a hard day's work with the tongs. Then, while they were resting and had nothing else to do, they walked back home. No doubt they were ready to eat their suppers when they got there, but that is a punishing day.

I would give a great deal to be able to turn time back and work with the old canoe fleet for a season. But, on the whole, I am very glad that my boat is equipped with a six-cylinder Studebaker.

The Racing Canoe

The log canoe was designed originally for work. But, since it was a fast boat anyway, it did not take boat builders and watermen long to see that, with a few alterations, the canoe could be made to go very swiftly indeed. All you had to do was to make the hull slimmer, the logs thinner, crowd on more sail, and you would have a racing boat instead of a working boat.

When these alterations were made, a major difficulty arose. The new craft were as fast as could reasonably be wished, but they were tippy or "tender." The problem was to find some means of gaining stability without losing speed. It was solved by the introduction of the "hike" or "hiking" board. This was a strong, smooth plank 12 to 15 feet in length and 10 inches wide. One end of the hiking board was placed under the lee washboard, while the other rested on top of the weather washboard and overhung the boat by several feet. If you put a few hefty men on the ends of these boards, you then have a perfect means of counterbalancing the pressure of the wind on a cloud of sail. It is perfect because your ballast is both intelligent and mobile. When the wind is strong, your hikemen slide to the extreme ends of the planks. When the wind weakens, they slide inboard toward the canoe's center of gravity. You want stout denim trousers for this kind of work, and a smooth plank, not a splintery one.

The hikeman must be absolutely devoted to his work, for on him depends the balance, hence the safety, of the racing canoe. In tacking he must look sharp when the boat comes about. It is his responsibility to shift the hike board to the new weather side and to scramble up on it for dear life, but not too quickly either. If the sails have not yet taken the full pressure of the wind, a half-dozen over-eager hikemen can turn the canoe over in a second.

On one occasion in a log canoe race in the Miles River, I saw the skipper of a lagging craft order his heaviest hikeman to jump overboard in order to lighten the boat. There were about two hundred pounds of excess ballast. The heaviest man on board immediately plunged over. He could not swim a stroke, but the boat was rounding a buoy at the time, and he managed to dog paddle up to it before drowning. There he clung until picked up by a race official's boat. It is of such mettle that successful hike-boardmen are made.

Canoe racing has had many ups and downs, and its history is by no means continuous. It is said that organized races were held at St. Michaels in the Miles River as early as 1840, with as many as thirty starters. A formal prize or cup was offered and a system of handicapping instituted. The contestants in these early races sailed their craft to a predesignated beach and there struck and furled all sail. They then pulled the sterns of the boats up on the sand, and the crews went ashore. At a signal, they raced for the boats, launched them, and made all sail as rapidly as possible. A standing start of this kind must have been something of a brawl. I should think the temptation to trip a rival crewman running for his craft must have been impossible to overcome. Still, it was a test not only of sailing ability but of seamanship in the smart handling of gear.

The Civil War ended formal canoe racing, after which it was revived from time to time, only to fall into disuse again. Finally, in 1924, the newly-formed Miles River Yacht Club revived the sport, which has become a yearly event in the Miles River Regatta ever since.

Most of the canoes which race in the event are old, built before I was born. The *Magic*, for example, which is moored in a cove at the village of Tunis Mills, was built at St. Michaels by Charles Tarr in 1894. The *Island Blossom* was built on Tilghman's Island by W. S. Covington in 1893. And so it goes. Only four racing canoes were built in the nineteen thirties. All other new additions to the fleet are old workboats which have been bought from watermen and rebuilt for racing. I can remember when the *Magic* had a gasoline engine in her.

But this supply is definitely limited. There are very few workboat canoes left. Since the art of making them has been lost, it seems only a matter of time before log canoe racing will become a memory.

The Bugeye

Anyone with an eye for boats who has ever seen a Chesapeake Bay bugeye is not likely to forget her. The tall, raking masts, the low free-board, and the fine sheer line are not easily mistakable.

After World War II, my wife and I bought a canoe and some supplies at Fort Ticonderoga on the lower Champlain and paddled in a leisurely way up to the Canadian border. For a while I had a little trouble with beaching the canoe at the end of a day's travel. I carefully carried her well ashore to keep the tide from setting her adrift during the night. My only excuse is that I was raised on salt water. After a while, I mastered the fact that there is no tide in Lake Champlain.

One day, in the midst of all that fresh water, I looked up from paddling and saw a bugeye at anchor, very snug and trim in a well-protected cove. I nearly fell out of the canoe. She was the *Little Jenny* out of Oxford, Maryland, and she was built in 1884. She was fifty-four feet long and all bugeye, one of the prettiest I have ever seen, though her lines were spoiled somewhat by too big a house amidships. Just looking at her made me homesick. The owner and his wife were very kind and showed us their craft from stem to stern.

A bugeye is so characteristic of Maryland waters it is unmistakable anywhere, even on Lake Champlain. And yet these vessels have had a very short history compared to that of the log canoe. The bugeye came into existence at the time of the Civil War because of an immediate and practical need: the necessity for a vessel big

19 LOG CANOE, SHOWING CLUB AND SPRIT SAIL PLAN

20 LOG CANOE, SHOWING CURVE OF WING LOGS, CENTER BOARD AND SHARP STERN

21 MODEL OF LOG CANOE

22 MODEL OF CHESAPEAKE BAY BUGEYE

and strong enough to drag a dredge over oyster bars. The log canoe, well adapted to tonging, was obviously too light a craft for dredging.

This need of a vessel for an explicit purpose affected the entire design of the bugeye, from hull to sail plan. The dredge which the bugeye was expected to pull was then, and is now, essentially an iron frame with a toothed scraper at the bottom of it. To the frame a bag of heavy netting was attached to receive shell and oysters turned up by the teeth of the scraper. When the dredge starts to fill, the boat pulling it must be large and strong enough to keep it traveling across the bottom. The two-masted bugeye with a length of from fifty to seventy-five feet was just the craft for the job. A schooner could undoubtedly have pulled the dredge, but her high bulwarks would have made her difficult to work, due to the fact that the dredge must be winched up to the deck every time it is emptied. The sheer line of the bugeye falls away from her head to a low freeboard from midship aft, thus eliminating that difficulty of a long haul over high sides.

The schooner has another disadvantage. It draws too much water. After the oystering season, bugeyes were always used in the summer for freighting, and many of their loads were taken aboard in shoal creeks, so the bugeye was built to draw little water. To this end she was also assisted by the use of a centerboard, rather than a fixed keel.

The trick in dredging is to drag at just the right speed and with the proper length of cable. Too much speed causes the dredge to hop or jump over the bottom; too little makes the dredge bury, or "bear too hard."

To meet such exacting demands, a very efficient sail plan is necessary. It was found that a plain jib and fore-and-aft sails at the fore and mainmast did the trick. This plan allowed for a minimum of gear aloft and was easily worked. Under normal circumstances, one man, the captain, could sail the vessel, thus leaving his crew free to handle the dredge gear and cull the catch.

In order to assure evenness of pull, sail makers cut their canvas in such a way as to prevent all slackness and bagging. In addition, the shape of the sail enabled it to spill wind rapidly. The combination of shape and of a taut sail meant that the vessel could maintain an even sailing trim and thus avoid rapid rolling and heeling that would cause the dredge to hop over the bottom as the cable tightened or slackened with the puffs of wind.

The very points which made the bugeye perfect for its job all made her a pleasure to sail. That is why I saw one in the middle of Lake Champlain.

Unlike the log canoe, the bugeye is still worked, for Maryland law prohibits the use of power for dredging. The result is that dredging today forms one of the last strongholds of sail in the United States. Racy skipjacks and graceful bugeyes sail every season and work specified grounds, some of which are unfit for tonging but well-adapted to the dredges.

In the 1957-58 oyster season, 78 sail craft worked in the waters of Dorchester, Somerset, and Talbot counties. An additional 67 dredged in the State waters of the Bay proper, making a total of 145 vessels in the fleet. Not all of these are bugeyes by any means. Many are skipjacks and sloops, others are hybrids of one sort or another. In fact, only about fifty bugeyes were under sail in 1938, and the number must be appreciably less by now. Since the last new bugeye was launched in 1918 and none are likely to be built again, it seems highly probable that this fine craft is sailing rapidly toward extinction.

accidents

8

Chances

It is a known fact that people who go on the water only occasionally and for pleasure suffer more accidents, and more serious accidents, than do watermen. It is marvelous, nevertheless, that pleasure boatmen have had so few serious accidents. Their tendency is to take chances where none are necessary. Watermen take as few chances as they have to; and even when forcing one's luck is necessary in order to make a living, some margin of safety is still allowed. Furthermore, watermen have a very alert weather sense and are not often caught unaware. They also have the advantage of knowing intimately the waters that they work.

The Calculated Risk

Watermen are most susceptible to accidents when they take calculated risks. They know well the narrow margin between a rational risk and forcing themselves and their boats over that line into the territory of foolhardiness. The water has very little use for misjudgments, and it will often kill a man for being foolish.

I have already mentioned the fishing party from Annapolis lost in December of 1960. They left port Sunday morning. At the time they were leaving on the Western Shore, I was standing on the Eastern Shore and looking at my boat hauled up on the ways in Tunis Mills. I had just put ice planks on her the day before, and I was disgusted. The tide was too low for launching. But I said to *Sammy*, "You stay there, old hoss; Monday won't be fit anyhow." The weather was so threatening you could taste it. Any waterman could have told you what that leaden sky and the sneaky cold draught from the northeast meant. And yet that party boat captain took a chance on calm seas that he hoped would stay that way. He took a calculated risk, but he miscalculated. He lost his margin of safety when he also lost his bearings in a howling snowstorm that drove six freighters aground in the Bay on the same day that he and his party perished. Two were picked up by a freighter, but the chances of survival after an hour's immersion in ice-cold water are virtually nil, and the poor fellows died of exposure.

But misjudgment is just as dangerous in summer as in winter. The tendency among watermen to stay out as long as possible in the face of threatening weather is understandable when, as you already know, so much time is lost every year. Two or three summers ago, three maninose rigs were working in Eastern Bay near Tilghman's Point when a big thunder squall built up over Kent Island. The men knew very well that, with the tide turning to flood, the chances were good that the squall would strike their area. But they hung on. One more bushel, just one more bushel.

They were very lucky indeed. When this vicious bit of weather ripped into them, all their rigs were swamped. The men climbed onto the cabin tops and clung there until the wind and seas washed their foundering boats into shoal water.

Sometimes one factor out of many in a given situation is ignored or forgotten, and a serious misjudgment is the result. On one bitter cold morning an energetic St. Michaels tonger studied the situation: plenty of sunlight, but very cold; a sea running strongly before a northwest breeze of perhaps twenty to twenty-five miles an hour. He figured that if he could get to a lee, say behind Tilghman's Point or up the Wye River by Bruff's Island, he could make a day. He decided to try it.

Rather than stay out in the cold wind and spray during his trip down the Miles River against the sea, he settled down in his cabin at a small wheel from which he could steer his boat. At first, the trip was not unusual. He had made worse. But after a while he became uneasy; some intuition sent a warning to his brain. Something was not just as it should be. For a while, he brushed the feeling aside; and then he noticed that his craft was behaving very oddly. She seemed sluggish, response to the rudder was "soft," and she had a tendency not to rise well with the sea. There was no water in her, he knew that. So it must be something else. By this time, she was definitely down at the bow and taking water instead of spray. When he came out of the cabin to see what was wrong, she was wallowing about like a pickle barrel, rather than an able boat. He saw why. Every spit of spray that had struck her was where it landed, frozen into lumps, bumps and ropes of ice.

If the tonger in question had not been alert, the consequences might well have been serious. He could have driven her under.

Neglect

Neglect of gear and boat often causes accidents. A man who does not keep his motor in good condition is looking for trouble. It is no fun at all to roll helplessly in a trough. The ability to leave, and leave at once, in the face of a hard, sudden blow is of the utmost importance as far as safety is concerned.

Captain Will Wharton took the floor boards out of his crabbing skiff one morning to locate a persistent leak in the bottom, and stepped ashore to get some cotton for caulking. When he jumped back in his boat, the entire bottom dropped out of her. He found himself in a peculiar position. He was standing on his boat bottom, but that portion of his craft was now four feet below the water level and resting on the mud. That left Captain Will standing in water up to his chest and encircled by the still intact and floating side planks and washboard.

Everyone in St. Michaels had told Captain Will that he ought to renail that old boat of his. He was now utterly convinced. Amusing, perhaps. But it might have happened off Deep Water Point.

Fog

Fog seldom causes serious accidents, but it can be very exasperating and tricky. Most tongers and crabbers seldom work in shipping lanes, so that the danger of being run down is minimized. But time is lost and maddening minor accidents occur. Trot-liners, blinded by fog, frequently ground their boats or foul their lines in grass. It is possible, of course, for tongers to work out compass courses to the various bars, but it requires some means of measuring revolutions of the wheel per minute to work out distances as well as directions, and we are rarely equipped with elaborate instrument panels. As a rule, such refinements are completely unnecessary and quite expensive. Since we always work in sight of land, we seldom carry compasses. In foggy weather, we generally grope and hope.

Some watermen have developed a keen sense of direction and are very successful in finding grounds in fog. Others, like myself, are often hopelessly confused. The best thing to do is to stop the motor, let the boat drift, and listen. Everything is dead quiet, except for the foghorn on Bloody Point, which booms hoarsely away off like a wounded sea monster. A sudden flight of ducks startles you with the creak and whistle of wings. A gull drifts out of the fog toward the boat, and he looks as big as a goose. But, after a while, you hear the chunking sound of oysters striking a culling board, and you know some tonger has made a lick and emptied his rakes. You locate the sound, start your motor, and move toward it. Something as big as a buy boat materializes in the mist, and a twelve-foot giant stands, apparently suspended over nothing, and he waves a couple of big timbers around. You have arrived. Where, for the moment, is of little consequence. You know you are on cultch, for tongers are all around you.

The eerie, distorting effect of fog often makes the problem of location very difficult. In Leeds Creek, I have turned in to shore to see where I was. Here is a little spit of sand and a tree sprawled over it that I have never seen in my life before, in spite of the fact that I have passed by it from time to time for some thirty years.

You remember Captain Will Roe as the skipper who carried a dog stick. When he left sail, Captain Will stayed on the water and for many years went crabbing out of St. Michaels. He kept some of his old nautical habits, including that of running compass courses. He had laid out one in particular from the harbor where he kept his boat to the harbor mouth, and thence to Maiden Point, where he frequently put his line out. From the center of St. Michaels harbor Captain Will had run a course on a big can buoy at the mouth of the harbor. A can buoy is made of riveted steel sections. It is hollow, of course, to assure floating, and so heavy it can not be lifted from the water without machinery.

Since Captain Maurice and Captain Will Roe often left the harbor at about the same time before dawn, they had talked about the possibility of being able to use the compass course in case of fog. One thick morning, they had their chance. Captain Maurice was for waiting, but Will Roe assured him there was nothing to worry about. All Maurice had to do was to stick close, and he would get them both out of the harbor and onto trot line depth at Maiden Point.

They started, each in his own boat and dead slow, of course, while Captain Will bent over the compass and steered the first leg of the course out to the can buoy. Captain Maurice ran along beside him not fifty feet away; but, even that close, it was hard to make out his friend's boat and he couldn't see Will at all. Captain Maurice grew more and more uneasy. It seemed to him to be taking an awful long time to run from the harbor to the buoy.

"Will!" he called, "Oh, Will! I ain't seen no can yet. Sure you're on course?"

A disembodied voice replied, "Certainly we're on course. Think I can't read a compass?"

At that moment there was a tremendous booming bong. Captain Maurice, from a position sprawled over the engine house, turned his head to starboard and watched the can buoy rumble down the side of his boat. He staggered back to his feet and called Will again.

"Will! Oh, Will! You was right on course!"

Again the disembodied voice replied. "Why, Maurice, I expect everybody in St. Michaels knows that by now."

Captain Will was good with a compass, no doubt of it, and Captain Maurice had a new piece of oak fitted in the stem of his boat to prove it.

The Will to Work

Accidents and injuries are sometimes caused by lapses of attention, momentary forgetfulness. Captain Billy Wales, for example, lifted a load of oysters in his rakes and took a step sternward to get a more comfortable position for emptying his rakes. He found there was no comfort at all in icy water, for he had stepped right overboard. His boots filled quickly, and it was a real struggle to gain his boat, hook one leg over the rudder, and so scramble aboard over the transom. He was getting on in years, anyway, so he figured he had had enough. He went home and sold his rakes.

Actually, however, it is the very will to work which causes watermen trouble. They drive themselves hard at times, and so sprained wrists and thumbs are a commonplace. The man from whom I bought my twenty-fours broke his back with them and has never been able to tong since. He had to crawl on his belly over the floor of his boat to reach his engine and start it. Steering was almost impossible. The Tidewater Police noticed the erratic wake and helped him into St. Michaels. And he will tell you himself that he ruined his back because, whenever he found a patch of oysters, he had to get all of them in one day. The problem of pacing himself at work is one that a waterman sometimes has difficulty in solving. And there is no doubt that his energy and determination keep him out in weather which is potentially dangerous.

The dredge fleet have the same problem. They have a comparatively short season, since they are not allowed to work until November. The men press themselves hard and sometimes crowd their luck too far. When a man from the fleet is lost overboard, the captains of the dredging vessels often find it difficult to get a crew for a while. But soon they are back again, and they are just as liable to the odd accident as other watermen. A few winters ago, a dredge boat struck a submerged object and sank within minutes. Fortunately, an oyster patrol boat was nearby and pulled the men from the water.

The Old Men

When all is said, it is the old men whom the water harvests most frequently. Captain Will Wharton lost more than the bottom of his boat two summers ago. I saw him the day before he died. He had gotten a bushel of crabs in that disastrous summer when a half bushel was a marvelous catch. He told me with that big voice of his he was going right back to the same ground the next day and get another bushel. But he never got that bushel. Another crabber, when he started to lay his line in Lloyd's

Creek, heard a boat engine. He looked around and finally located the boat in the duck grass, its motor still turning tirelessly. It was Captain Will's crabbing skiff, and it was empty. His body was found later. Presumably, his heart had failed him.

The man who found the boat says that Captain Will's trot line was laid as straight as a string, a good, workmanlike job on the last he ever did. No doubt, if he had stayed ashore, Captain Will might have lasted longer, but he would have been miserable. The eagerness of watermen to go out, and their notorious crankiness and misery when kept ashore, partly account for the large numbers of old men on the waters. In one boat alone, I have seen three tongers working together whose aggregate age was 245 years. Their craft had a load of years to bear, but the old-timers, in good weather, frequently put out twenty-five bushels of oysters.

Sometimes men, but more often women, will complain about old men going out on the water. The women will say, "I don't see how old Cap'n Williams stands on those washboards without someone holding him up. Those old men oughtn't to be allowed to oyster. Cap'n Williams is going to fall overboard one of these days, and that will be the end of him."

Of course, some of them do end their days in just that way. In January of 1961, a Tilghman's Island waterman was drowned in Knapp's Narrows. A passing buy boat swamped his small craft with her wash. The man lost was eighty years of age.

To call such an accident to an old man a "tragic incident" is to use the jargon of magazine articles and radio broadcasts. Those are the chances you take, and, if you are old, the chances are much against you. It figures.

I remember some years ago one of the first watermen in our area to trot-line for crabs was lost near the head of the Miles River. He had complained that crabs were scarce and said he was "going hunting"; that is, he was going to locate new ground where crabs were more plentiful. I like the spirit of a man seventy-eight years old hunting for new grounds. He never found them. In leaning over to clear grass from his wheel, he turned dizzy, lost his balance, and fell overboard.

Watermen are accustomed to handling problems alone, day in and day out; and so, when a man drowns with whom you have crabbed and oystered for years, you notify the Tidewater Fisheries Patrol, and then you help to hunt for the drowned man yourself. It is the last thing you will ever do for him.

The fleet takes up anchor, stows gear, stops work. One crabber or one oysterman hails another, and so the news is passed. The boats move from every quarter up the river where the old man's boat was found, empty and dead on the water. The very grapnels which were used to anchor boats or lines are now employed in a grimmer task. The men do not stop until the job is done. Their strange fishing completed, they tow him home in his own boat, and they say he was a good man in his time, a good man with the trot line, a good man with the tongs, but too old for the life. They feel old themselves, many of them, too old to follow the water. But the next day sees them back at the lines, back at the tongs, a rough summer sea leaving a crust of salt on their jackets, or a winter sky spitting a few flakes of snow at them just to show them what the life is and how they can expect to leave it, once and for all.

To those who say that the old men should not be allowed to follow the water, it is of no use to point out that the crabber who drowned lived his life at the job he liked best, that he died cleanly and full of years. It is the general feeling among landsmen that a man should live as long as he can, whether or not he is of any joy or use to himself or anyone else. The old crabber died suddenly, that is true, but he was active and interested, and his fate was far better than the slow dry rot of aged and empty years. Let them oyster as long as they can stand on the boards and lift the tongs, let them crab until the dipnet falls from their hands. It is better than an armchair and twelve ounces of tobacco a week.

I have seen a man in his seventies row a heavy, eighteen-foot skiff from St. Michaels harbor, tong for six or seven hours, row his skiff back to the harbor, shovel his oysters, and put out six or eight bushels. He was useful, he still had his pride. He was a man yet, and he knew it.

the character of watermen

9

The Dodo

I hope that what I have already written has dealt, at least indirectly, with what watermen are like, their general characteristics. But it may not be inappropriate to sum up some salient points and to add a few details. What I have to say may also include the additional flavoring of nostalgia. A friend of mine, also a waterman, has told me that perhaps someone should attempt to tell people what free watermen are like, since it is entirely possible that they may soon disappear altogether. It is as if some naturalist had seen and recorded the habits and characteristics of the dodo just before its extinction. I do not take so gloomy a view, myself. But perhaps my friend is right, and I should follow the practice usual in bird listings of an extinct species and put a black band of mourning around the chapter headings.

However, these men are still very much alive, and are consequently difficult to pin down.

Many of my friends and acquaintances who have led sedentary lives in the larger urban areas have often been struck by the large number of older men engaged in one or the other of the commercial fisheries of the Bay. They wonder that men who have not in all of their long lives spared themselves hard work should still be alive at all, let alone active and able.

First of all, a certain amount of physical toughness is necessary. Men having good constitutions are likely to wear well in any way of life, and it is at once obvious that no man with a fundamental physical defect, such as heart trouble, need apply for a job as a waterman. On the other hand, it is a mistake to suppose that a fit man who uses common sense on the job will shorten his active life by continuous physical exertion. In fact, the opposite would seem to be true.

We are animals possessing fairly powerful bodies and extensive musculature. The conditions of our lives are such that millions of people rarely have any real need to use their bodies, and those conditions will almost certainly be multiplied and extended as technological devices and know-how are applied increasingly to every form of useful and useless endeavor. We know the nervous fatigue of paper work done under tension and pressure; we know the exasperated fatigue of boredom in the endless repetition of small and identical tasks; but millions of us seldom experience the relaxing, grateful and healthy fatigue of a used (not over-used) body.

Many years ago a friend of mine, an able doctor with an active mind, studied diseases of the heart over a long period. He came to some interesting conclusions not generally known at the time, but since widely publicized. One was that heart fatalities occurred mostly among men in their forties, a period when they should have been near the peak of their physical prime. The other observation he made was that far the greater number of fatalities occurred among men leading completely sedentary lives. Other less dramatic but still serious disorders have been traced to the same condition. I realize that cause is seldom single, but generally multiple. Still, we may be paying a heavy tax for static lives, and the price will be higher as the years go by and there is less and less opportunity to use the excellent equipment provided by nature for an animal which no longer needs it.

Up to the beginning of the twentieth century, all normal adult men and women had a natural outlet for physical energy in their daily lives: they walked. Now, almost no one walks. I have rambled for hours through large suburbs in Baltimore and Philadelphia without ever seeing another man or woman on foot. I was the single adult pedestrian. Only children showed any need or desire to use their legs. In many sections, sidewalks were allowed to deteriorate, since there was no use for them any more.

So little are we accustomed to physical activity, that in many cases it is dangerous for men to discover that they have limbs after having forgotten the fact for twenty years or so. I know of a man who suddenly remembered. He was delighted. After twenty years of a completely sedentary existence, he discovered what very good fun it is to walk mountain trails. He died of heart failure.

Watermen are not health fadists, nor have they any attachment to hard work for its own sake. That is not part of their character. A romantic fondness for quaint inconveniences would strike them as being on the edge of insanity. Few people have a greater appreciation of small comforts than those who have none every workday. Six hours of tonging in a bleak, raw wind, a runninng sea, and a tide that whips the shafts like switches, give a man a very keen and grateful sense of just what a house is—a place of shelter. Most of us have forgotten this fundamental definition. Linemen, surveyors in the field, bridge and dam workers, all of these men know what a house is.

I do not believe that continuous exposure to bad weather and arduous physical labor in themselves make men better. I do believe that the various jobs done by watermen and the conditions of their existence bring out and develop latent characteristics which become a part of the men themselves. But men who must deal frequently with adverse outdoor conditions are made much better by having warm, cheerful, comfortable, and adequate shelter when the work is done for the day. They have neither the need nor the desire to "rough it."

Watermen believe that those people who beat their hairy chests and then break ice in midwinter to go swimming among the floes are either show-offs or plain damn fools. Nor is a tonger at all likely to play basketball on the weekends. He is far more likely to sit down in a duck blind and wait for the birds to come to him. What he wants is sedentary diversion and rest. In short, these men are not hard primitivists. That is strictly a romantic idea that entertains those who have never had to do hard physical work in bad weather.

I have frequently associated with people who have a highly developed aesthetic sense. They are capable of taking very keen pleasure in the out-of-doors from the standpoint of seeing and appreciating its beauty. I have often been asked just how much watermen, who see nature constantly in all her variety of moods, are affected, aesthetically, by what they see. It is a question whose answer throws some light on the character of watermen. What I have to say about it can only be in general. We may have among us mute, inglorious Miltons, for all that I know; the difficulty is that, if they exist, they are mute, and therefore difficult to distinguish.

However, my impression is that most of them are like Wordsworth's Peter Bell:

> A primrose by a river's brim
> A yellow primrose was to him,
> And it was nothing more.

Actually, these men, like Peter Bell, are very practical and tough-minded. Refined sensibilities and trained aesthetic perception, like other accomplishments, take leisure and learning to develop. They haven't the time for it, nor are they usually in contact with those who have. It is probable that a "certain slant of light on winter afternoons" falls upon unseeing eyes. I think they seldom consciously enjoy the beauties of nature, unless these are very obvious and striking, as in rainbows. Their response to a lovely day is physical rather than aesthetic. It can send them singing to work, bursting with energy, high spirits, and a general sense of well-being.

This is not to say that they see little. They see far

more, and in far more detail, than most of us ever do. They are constantly aware of, and alert to, their surroundings. But their interest is not aesthetic. Captain George Roe looks carefully at the sky in the west, studies the movement of a bit of seaweed floating past his boat, thus gauging the tide, turns and scans a cloud bank in the south, and announces to anyone who wishes to listen that the wind will haul southerly in approximately fifteen minutes. Nearby tongers on the bar nod and check their watches. In about twenty minutes, the wind hauls and blows southerly. A feat of this kind shows remarkable knowledge and a highly critical appraisal of natural surroundings, but it can scarcely be called aesthetic.

What watermen remember best and admire most is absolutely the worst that nature can show, rather than the best. They deeply respect a most worthy opponent.

"Cap'n, when I crossed that river, she was smoking. The spray went clean over Oak Creek bridge, and they found salt crystals on the tree leaves a hundred yards from shore."

And they shake their heads and smile, part glee, part rueful awe at the sheer, dirty wickedness of a blow like that, the bastard.

However, they constantly watch the life that surrounds them. They mark every flock of geese, they see swans feeding on the flats and hear their calls. There is scarcely anything that moves in the air, on the water, or under the water that they do not see and remember.

In the fall when numbers of loons from the north stop in the Bay and its tributaries and fish our waters, they are noted at once. The loons calling to one another on a quiet day can set the whole fleet off, answering their eerie, musical cries.

Among these many wildfowl, the horned grebe or hell-diver is a favorite with oystermen. We call them pigwitches. They are little witches, all right, but so far as I know they and pigs have never met. They feed by diving and they will line up, four or five at a time, and wait alertly for the tonger to stir up the bottom with his rakes. They dive, then, and catch whatever his tongs have disturbed, small fish, blood worms, and the like. Most tongers are fond of pigwitches. These sleek little birds are often company on a lonely job, and many a tonger who has had little luck has jokingly complained of working all day for the pigwitches.

Watermen will remember a great school of blues seen thirty years ago, and the glimpse of an eagle harassing ducks will never be forgotten, a story often repeated.

They are not professional nature lovers or bird-watchers, but they are, most of them, very keenly aware of their surroundings.

Honest and God-fearing

Mr. Byrd, the chairman of the Tidewater Fisheries Commission, likes to address his watermen as "honest and God-fearing men." A statement like this, vaguely flattering to those so addressed, and no doubt paternally and kindly meant on Mr. Byrd's part, yet shows how difficult it is to generalize about any given group of people.

As for God-fearing, most watermen "believe in the Old Man." But the stiff-necked, pious, hell-fire, Bible-quoting waterman belongs to an earlier generation. One can lament his passing, but my distinct impression is that he is passing. Nowadays, most watermen appear content to send their children to Sunday school and their wives to church, while they themselves rest after a hard week with the tongs, or bait their crab lines in preparation for Monday.

Honesty is another matter and requires a closer look at the waterman through eyes that are not his own.

First of all, men who spend all of their lives in settled occupations, who will do tomorrow what they did today, who work steadily at a given task, live in a regular and familiar routine: these solid, substantial, rather dull and respectable folk are likely to be critical of those whose lives seem incredibly chancy, unsettled, and harum-scarum, like those of watermen.

But, in addition to traditional attitudes of this kind, gradually, during my lifetime, a pejorative connotation for the word "waterman" has slowly accreted in the collective minds of many communities. The precise pattern is often not clearly defined, but the general outline is plain to see.

First of all, watermen are "shiftless." They come off the water and spend their money every night on beer. Easy come, easy go. Tong the dollars from the bottom you have never sown, reap the harvest of crabs you have never cultivated, and spend it all in one night. Tomorrow there will be more. If not tomorrow, why then the day after.

In order to "prove" this point, it is necessary only to single out a couple of watermen notorious for wasting. It is easy to forget that most watermen do not waste, and that some landsmen in regular occupations do.

Secondly, watermen are "light-fingered." They will steal anything on the water, including your illegal crab pot. If you do not watch them, they will fish your gill net. They will even lift one another's eel pots and/or the contents thereof.

And yet the workers who built the Chesapeake Bay bridge, most of them farmers, carpenters, and the like, walked off the job with loot that would have made a waterman look like a piker. Such evidence, however, is conveniently ignored. There are honest and dishonest watermen, just as there are honest and dishonest merchants or farmers.

Thirdly, watermen are vaguely "low class." What low class consists of precisely is of no importance. It is the kind of epithet which explains itself: that is, it explains everything and nothing, since it is never defined.

What makes them low class? On the Eastern Shore of Maryland, to work with your hands, skillfully and with knowledge, is not considered to be dishonorable.

Nor is it a matter of education or of language. In the United States, there are two dialects of English. One is the official language of the schools, and the other is the vulgar tongue. One is half-learned and forgotten, and the other is used daily in speech. Watermen are no different in this respect from others. They use the popular or vulgar tongue.

Employers of men have frequently told me that they preferred not to hire watermen. They say all these men can do is fish, crab, and tong; nothing else. They are "low class" workers. This criticism, as it stands, is untrue. The best boat carpenter I ever saw in my life used to down tools the first of September and pick up the rakes. Nothing in the world could keep him ashore, including the entreaties of his boss. Watermen help to build and paint half the houses in Talbot County, and most of them are able at a variety of jobs, from boat repair to tree felling and trimming. Some watermen have become so skilled at land jobs that they "can't afford" to go back to the water.

No. What employers actually object to is something different. For one thing, many watermen are actively unhappy ashore. They work hard and well, but the time comes when they can stand it no longer. So an employer may be looking at a workman one minute, and at thin air the next. He has gone, and sometimes without a word of either explanation or apology.

Furthermore, watermen are not docile. They have an inveterate desire to do things their own way, just as they do when they work on the water. Nor do they at all like taking orders. A nagging boss will soon find himself minus a workman. If these characteristics make him a "low class" worker in the eyes of an employer, they do not lower him in the regard of other workmen, many of whom are equally capable of "telling the boss to go to hell."

It is difficult to pin down what, after all, is so ill-defined, but I think the greatest stronghold is among women. Women usually want for their sons "something better" and for their daughters anything but a waterman for a husband. If there is one thing in the world that a woman dislikes, it is economic uncertainty, and the waterman's life is always a gamble. Year in and year out, year after year, he makes a living, sometimes a fairly good one. But his pay-check does not arrive punctually every week in the year with a known amount which can be depended upon. What can a woman do about a week or two of ice which has kept her tonging husband home and no money coming in? No welfare for him; he is, in the eyes of the law, an independent business man. It is true that the week before the ice came, he had done well, but to his wife's mind that is beside the point. What she thinks about, and that alone, is this week, this week of ice.

He is just about to tong again, gets a few days in, when a northeaster, accompanied by sleet and freezing rain, pins him down again. After which, the wind hauls northwest, blows like hell, and freezes the water into iron. His wife is now on the point of using a blowtorch, either on the ice or on her husband, preferably the latter. And she swears a great oath no daughter of hers will ever marry a low-class waterman.

Or a crabber dips five bushels of jumbos and a barrel of culls each day for most of June. And then for most of July, he wastes bait and catches little or nothing. He tells his wife August will be better. She tells him to get a job on land and stay there, and she is going to make sure no boy of hers ever does such a crazy thing as to follow the water.

But nothing helps. Her husband continues on the water. Her son, as soon as he can walk, may very well take to it like a duck for the rest of his life; and the young watermen continue to marry.

Culling Honesty

While on the subject, it may be well to examine a question which has been much in the mind of Mr. Byrd, and has often occupied the best energies of his Tidewater Patrol. How honest are tongers in their culling?

Generally speaking, tongers realize that next year's harvest depends upon leaving behind on the grounds this year's small oysters, and they behave accordingly. In this rational and intelligent behavior, they are notably assisted by the police. It is a hard thing to tong and cull honestly, and to watch a man next to you throw in everything he picks up: shell, small oysters, stones, and boxes.[1] It is only the sure knowledge that the police will get such a man, if not this week, then next, that keeps the honest tonger steady. And the police do get him, generally sooner than later. They know and watch carefully the few men who deliberately violate the culling law.

The tonger is allowed a leeway of five per cent of "trash" (shell, small oysters, etc.) per bushel of market oysters. Some trash is inevitable; otherwise the process of culling would take so much time there would be little left for tonging.

What gets most tongers into trouble is carelessness due to hurry. Every minute saved at the culling board means one more for tonging, and the more tonging you do, the more oysters you catch. And remember, the culler has literally hundreds of judgments of size to make, and to make as rapidly as possible. He can not stop to measure every oyster.

Three tongers, an expert boat carpenter, a first-rate boatyard worker and tree feller, and a house painter and captain of a large pleasure cruiser worked one day together in the same boat on Ashcraft Bar. Now, this bar is notorious for its heavy production of small oysters, so it always presents a culling problem. Much time is consumed in sorting out market oysters from a mass of small ones. As the day wore on and time pressed, the culling grew more and more careless.

Late in the afternoon, they decided to quit for the day, and they ran into St. Michaels harbor to the oyster house, where they sold their jag and shoved away from the bulkhead. As they started their motor and steered for

1. A box is a former oyster whose two shells are still joined together. When struck by a culling hammer, it makes a hollow sound, like that of an empty box.

the Inner Harbor, a Tidewater Fisheries policeman walked out on the landing of the oyster house and began to inspect the oysters just landed.

Legally, the three tongers were in the clear. They had sold their jag. During or before sale, the tonger is responsible for an illegal percentage of trash; after sale, the buyer is responsible, and must pay any fines incurred.

These men immediately turned the boat around and ran back to the oyster house where they awaited the completion of the inspection. When the inspector announced over five per cent of trash present in the jag, they assumed full responsibility and paid, among them, eighty-four dollars and thirty cents in fines.

I have always respected these three tongers for their manly behavior in turning back to the wharf; but it also demonstrates how expensive carelessness can be.

In the case of obvious and deliberate violation of the cull law, not only is the tonger fined, but he is required to shovel back overboard his entire jag. He loses a day's work and pays for it. I have always figured, myself, that if I couldn't make a living honestly, I'd have to try some other trade, and the police are able supporters of that view.

Self Reliance

Watermen are the most opinionated, cocksure, critical, conservative, and independent group of men I have ever encountered. Nothing you do will entirely please them; and nothing you do will visibly impress. They may very well be the last living specimens of an almost extinct species, except in political theory: the independent, the individual man.

No country, and no political theory, has ever evolved a means of dealing with large masses of men as individuals; men in great masses quickly degenerate into numbers for all practical purposes of government, including the vote. It may be that it is quite impossible to deal with hundreds of millions of men and women as individuals rather than numbers. I do not know. All that I do know is that the independent man is a fossil, except for a few isolated groups, such as the watermen.

With them, their own choice, their own opinion, is openly of paramount importance. Servile good citizenship and the spirit of cooperation have not yet sapped their manhood. In fact, watermen are distinctly leery of anyone who wants them to cooperate. They are so backward that they believe cooperation often means, "Now, you do what *I* want, and *we* will get along fine together." I know it is a shocking thing to report, but truth compels me to say that they have an invincible prejudice for doing what *they* want to do.

I think it is significant that no mail-order house has yet succeeded in selling them their boats and their gear ready made. Either they make their own boats and their own gear, or they go to an independent craftsman who makes them himself. And in the matter of boats and gear, there are as many independent and irreconcilable opinions as there are watermen.

Independence is fostered by the job itself. A life lived mostly alone on wide stretches of water under all conditions of sea and weather creates self-dependence impossible to masses of men crammed into cities, exurbs, and suburbs. If an oysterman's motor catches on fire, he can not call the fire department. He must cope with the matter himself, and it is likely to be one hell of a long swim if he doesn't. If a crabber comes upon another man stealing his eel pot, he deals with the situation himself. If his boat goes dead in a gale and wallows dangerously in a heavy trough, he must be able to handle the crisis or drown. If his gear breaks, he mends it or loses time at work. He can not draw new tools from the storeroom clerk.

At the same time, he learns to endure. A tonger, his hands always wet, sometimes has to walk up and down in his boat and beat the cold-ache out of his fingers before he can go back to work. And he goes back to work.

The crabber, from constantly salting and handling salted baits, has open nicks, scratches, and cuts on his hands that never properly heal. Salt in such lesions is literally hell. But he works his line, and he salts his bait. Every able waterman develops self-reliance and dogged determination as qualities inseparable from his character.

By comparison, city dwellers and those housed in the dreary congeries of homes called suburbs are pitiable in their helplessness and lack of self-reliance. It is not their fault. They are served on the one hand, and actively discouraged from any self-help on the other. When, some years ago, a severe water shortage threatened New York City and its teeming appendages, only one man of twelve million did anything about it. He drove a well in his garage and was promptly arrested. The independent man can no longer be tolerated in our complex and crowded society. He gets in everyone's way, in the manner of a crowbill that wishes to swim one way while a thousand others are going cooperatively in another direction. He is at once trampled under foot. He is a nuisance.

If you find that you have grown a little weary of falling down a slippery pair of steps in front of your house, forget about it or employ a railing corporation to "help" you. Do not try to put in a railing for yourself. You will be discouraged. In fact, if you have not taken a full day or so from work to go through the complexities and details of getting a permit (you are generally regarded as either lunatic or having communist leanings), you will not only be discouraged, you will be arrested.

If you object to paying blackmail prices to electricians for a little wiring, do not think of raising so much as a little finger in the direction of doing the work yourself. Should your house burn down because of a live cigarette left by a meter-reader on the front porch, no insurance company will give you a cent.

Such an environment does not make for confident self-reliance. It puts a heavy tax on such qualities. Those characteristics which make the handling and management of sheep easy and agreeable are the ones at a premium in large communities. I have seen a whole block full of people thrown into mindless panic because the turning of a faucet produced no water. The stores were open. Beer is far better than water anyhow. There is also something, though not very much, to be said for milk.

The Law

Generally speaking, watermen are not law haters. They are law interpreters. If they find a law to be good and reasonable, they will abide by it, though not to the foolish letter. These men do not take easily to a harsh, arbitrary, and foolishly rigorous application of the law.

Understanding and good judgment have been proved of real necessity on the part of enforcement officers. It is best to be firm, yet reasonable; just, but not offensive. An overbearing and inexperienced law enforcement officer can cause trouble, perhaps grave trouble, for himself and others.

I have often been amazed at the amount of abuse handed out by city policemen to both pedestrians and car drivers for the most minor infractions of law. On the water, such an officer would not last a week without finding himself overboard or flat on his back.

Above all, watermen like one of their own to enforce the law. In the first place, watermen distrust foreigners; that is, any man not born and bred on the Eastern Shore. In the second place, a shoreman, preferably one who is a resident of the same county where he works, will know local conditions and local men. He will be in a better position to act efficiently, yet without stupid, letter-of-the-law, blanket blindness.

For example, it is neither necessary nor desirable to arrest an oysterman because he has not procured his tonging license within thirty days of the opening of the season. He may well have forgotten the matter completely, without any malicious intent to defraud the State. A courteous but firm reminder is sufficient. If it is not, then is the time to apply pressure.

Again, a man who is guilty for the first time in a season of having a little over five per cent of trash in his jag of oysters is not deliberately violating the law. A warning in such cases is enough.

On the other hand, a policeman who winks at trash one day and then slaps a fine for the same offense the next will lose all trust and all respect. Such an officer can do more damage to the intent of the law and its efficient enforcement than all of the violators put together.

I have said it before, and I will repeat it now, we are fortunate in the patrol officers who have for years inspected the fleet in the Miles River and Eastern Bay. They keep clearly in mind the intent of the law, which is not primarily to fine violators but to conserve the resources of the Bay.

Manners

Watermen in general willingly help one another. This spirit extends even to pleasure craft and pleasure boatmen, most of whom, be it said to their credit, stand by when aid is needed on the water. Water is an unstable and frequently dangerous element where help may at any time be needed, a fact even more pressing and obvious in the case of professional watermen who must earn their living in all weathers. There is an unspoken bond of hardships endured and dangers overcome in a common trade, even though rivalry among individuals and groups may run high.

In cities, a single service or a single article wanted at the same time by more than a dozen people will be very likely to cause rudeness and shoving. Even after a hard day's work in trying weather, it is rare to witness the slightest jostling or rudeness among oystermen with their craft clustered around the hull of a buy boat. To the contrary, they patiently wait their turn and help one another. I have had many a bushel of oysters hoisted by men who already had their hands full with their own.

I do not mean that these men are models of good behavior. They chuck a good deal of rough and bawdy talk around, and they sometimes demand excessive and violent pleasure after hard and wearing work. Also, it can be dangerous to press a waterman too far, for his reactions are sometimes very sudden by usual standards.

A tonger whom I will call Henry found a nice patch of oysters out in Eastern Bay. They came up well–big, clean oysters, and Henry was very happy about that. He had been looking for a patch for several hours. However, he was not left to enjoy his patch alone. In a short time, he was joined by another tonger. This is an accepted practice. After all, you tong oysters wherever they are. It is not considered good manners, however, to anchor close and sheer your boat in such a way as to drive a man off his patch and then tong it for yourself. But that is exactly what happened to Henry.

He got down off the washboards, hung up the pair of rakes he was using, and walked aft to his anchor line. He looked at the man who had sheered him off his patch, and he said, "Cap'n, I'm about to pull my line in and take up my anchor. I'm going to start my motor and make a little circle out there with my boat. I'll take my time. But if you are still here when I make that circle, I'm going to cut you in half. And I hope there won't be no hard feelings."

The man looked at Henry. He said, "I believe you mean it."

And Henry replied, "There is one sure way to find out."

By the time Henry got his anchor up, he no longer had company.

I will not say that all watermen are entirely of Henry's mettle, but they possess a considerable share of it, just the same. Therefore, amiability and a desire not to behave in such a way as to cause others unnecessary troubles are qualities it pays everyone to encourage.

Social Position

The Talbot County village of Tunis Mills where I live is not a waterman's settlement. There was a time when twenty to twenty-five boats worked out of Tunis Mills, but that is no longer true. I am now the only active waterman living here. The inhabitants have seldom been watermen by choice.

Nevertheless, in its beginning, this community, like so many others on the Eastern Shore, was closely associated with the water. A large sawmill was established on a point of land near the head of Leeds Creek and wharves to deep water were built around it. Great rafts of logs were towed from all directions to the mouth of the creek and up to the Mill Point for sawing. All of that was before my time; the sawmill has disappeared, and the wharves are rotted to stumps of pilings. The exhaustion of large stands of timber near water and so easily transported by raft meant the death of the mill. Many of the mill hands, loggers, sawyers, foremen, and the like, left with the closing of the mill. Others remained, and they and their children, their children's children, now form part of the population of the village.

Only one other industry held on, and that was a canning house on the other side of the creek from the mill. It continued to operate until I was in my 'teens. Very few of the village men worked there, the hands being mostly colored. I can also remember swimming at the bridge in an ebbing tide and being festooned with tomato skins, for the cannery pumped all its refuse into the creek. The cannery was established after the failure of the sawmill.

It was a curious thing to see how the social desirability of certain portions of a community finds its way even into such a tiny place as Tunis Mills. The older people were firmly convinced that we lived on the "wrong side" of the bridge. They had learned this distinction when they were young and the mill was in full operation. It is a common prejudice that the further away one is from his place of employment, the more respectable his status in the community. The dividing line in our case was Leeds Creek. We lived on the mill side, the "wrong side." In my day, the distinction was meaningless, since the mill had long disappeared, and yet the older people continued to cling to the idea. The cannery, ironically enough, was on the "right side" of the bridge. We are now blessedly without either mill or cannery.

In a waterman's community, there is little distinction as to a more or less "respectable" geographical position. Where most of the adult male population works in boats, the nearer the place of residence to water the better. Such men have no desire at all to get as far as possible from their jobs. As for social gradation, either you are an able and successful waterman, or boat builder and shaft maker, or you are mediocre. You may even be a lazy, good-for-nothing waterman. These are the usual gradations.

The possession of money in the bank, the ability to pay

your own way, this is highly regarded. But the regard is placed where it should be, on the ability of the individual, on his pluck, determination, and capacity for hard work.

Watermen have their faults, and I have made no effort to disguise them. They are sometimes ignorant and superstitious, stubborn and lacking in social warmth. Often, too, they are narrowly provincial. I remember the case of a waterman who sold a stranger a peeler for bait. He charged him one dollar for the peeler, an outrageous price. Someone sidled up to the waterman afterwards and said, " Know who that feller was?"

"Never seen him in my life before; ain't from around here."

"You will see him, though. He's the new preacher."

That waterman faced the music. Next Sunday, he put on his best and went to church. After listening to an acid do-unto-others sermon, he made his apologies. But I don't suppose he likes strangers any better than he ever did.

Yes, they have their faults. But they are a hardy and independent people who tend to subject everything to their own opinions and their own experience. Such people are not easily governed or readily persuaded, but, take them for all in all, they are men.

When I hang up the rakes for good, I hope that young and vigorous men, stiff-necked and independent, will continue to follow the water.